RECOMMENDATIONS FOR ENHANCING REACTOR SAFETY IN THE 21ST CENTURY

THE NEAR-TERM TASK FORCE

REVIEW OF INSIGHTS FROM THE

FUKUSHIMA DAI-ICHI ACCIDENT

U.S.NRC

United States Nuclear Regulatory Commission

Protecting People and the Environment

This page intentionally left blank

RECOMMENDATIONS FOR ENHANCING REACTOR SAFETY IN THE 21ST CENTURY

THE NEAR-TERM TASK FORCE REVIEW OF INSIGHTS FROM THE FUKUSHIMA DAI-ICHI ACCIDENT

JULY 12, 2011

Dr. Charles Miller

Amy Cubbage

Daniel Dorman

Jack Grobe

Gary Holahan

Nathan Sanfilippo

This page intentionally left blank

DEDICATION

This report is dedicated to the people of Japan and especially to those who have responded heroically to the nuclear accident at Fukushima. It is the fervent hope of the Near-Term Task Force that their hardships and losses may never be repeated.

Throughout its tenure, the Near-Term Task Force has been inspired by the strength and resilience of the Japanese people in the face of the inconceivable losses of family and property inflicted by the Great East Japan Earthquake and tsunami of 2011 and exacerbated by the ongoing radioactive releases from the Fukushima Dai-ichi Nuclear Power Plant. The heroes of Fukushima shouldered the emotional impacts of the devastation around them and labored on in the dark, through the rubble, with increasing levels of radiation and contamination. They undertook great efforts to obtain power and cooling to prevent the unthinkable from occurring. The outcome—no fatalities and the expectation of no significant radiological health effects—is a tribute to their efforts, their valor, and their resolve. It is our strong desire and our goal to take the necessary steps to assure that the result of our labors will help prevent the need for a repetition of theirs.

ACKNOWLEDGMENT

Over the past 4 months, the Near-Term Task Force has devoted many long hours and an extensive amount of thought, deliberation, and collaboration to produce this report. The recommendations presented herein will likely require the same by those who will make decisions as to whether they should be endorsed or not. That is as it should be. The Task Force members comprise over 135 years of regulatory experience as a collective body. I have valued greatly their wisdom and dedication in completing this task. It has been my privilege to serve as their leader. So to:

Amy Cubbage
Cynthia Davidson
Dan Dorman
Jack Grobe
Gary Holahan
Nathan Sanfilippo

I say thank you,
Dr. Charles L. Miller

CONTENTS

All photographs and graphic images used in this document are either in the public domain, have been licensed for public use, or have no known copyright restrictions.

This page intentionally left blank

EXECUTIVE SUMMARY

The Near-Term Task Force was established in response to Commission direction to conduct a systematic and methodical review of U.S. Nuclear Regulatory Commission processes and regulations to determine whether the agency should make additional improvements to its regulatory system and to make recommendations to the Commission for its policy direction, in light of the accident at the Fukushima Dai-ichi Nuclear Power Plant. The Task Force appreciates that an accident involving core damage and uncontrolled release of radioactive material to the environment, even one without significant health consequences, is inherently unacceptable. The Task Force also recognizes that there likely will be more than 100 nuclear power plants operating throughout the United States for decades to come. The Task Force developed its recommendations in full recognition of this environment.

In examining the Fukushima Dai-ichi accident for insights for reactors in the United States, the Task Force addressed protecting against accidents resulting from natural phenomena, mitigating the consequences of such accidents, and ensuring emergency preparedness.

The accident in Japan was caused by a natural event (i.e., tsunami) which was far more severe than the design basis for the Fukushima Dai-ichi Nuclear Power Plant. As part of its undertaking, the Task Force studied the manner in which the NRC has historically required protection from natural phenomena and how the NRC has addressed events that exceed the current design basis for plants in the United States.

In general, the Task Force found that the current NRC regulatory approach includes:

- requirements for design-basis events with protection and mitigation features controlled through specific regulations or the general design criteria (Title 10 of the *Code of Federal Regulations* (10 CFR) Part 50, "Domestic Licensing of Production and Utilization Facilities," Appendix A, "General Design Criteria for Nuclear Power Plants")

- requirements for some "beyond-design-basis" events through specific regulations (e.g., station blackout, large fires, and explosions)

- voluntary industry initiatives to address severe accident features, strategies, and guidelines for operating reactors

This regulatory approach, established and supplemented piece-by-piece over the decades, has addressed many safety concerns and issues, using the best information and techniques available at the time. The result is a patchwork of regulatory requirements and other safety initiatives, all important, but not all given equivalent consideration and treatment by licensees or during NRC technical review and inspection. Consistent with the NRC's organizational value of excellence, the Task Force believes that improving the NRC's regulatory framework is an appropriate, realistic, and achievable goal.

The current regulatory approach, and more importantly, the resultant plant capabilities allow the Task Force to conclude that a sequence of events like the Fukushima accident is unlikely to occur in the United States and some appropriate mitigation measures have been implemented, reducing the likelihood of core damage and radiological releases. Therefore, continued operation and continued licensing activities do not pose an imminent risk to public health and safety.

However, the Task Force also concludes that a more balanced application of the Commission's defense-in-depth philosophy using risk insights would provide an enhanced regulatory framework that is logical, systematic, coherent, and better understood. Such a framework would support appropriate requirements for increased capability to address events of low likelihood and high consequence, thus significantly enhancing safety. Excellence in regulation demands that the Task Force provide the Commission with its best insights and vision for an improved regulatory framework.

The Task Force finds that the Commission's longstanding defense-in-depth philosophy, supported and modified as necessary by state-of-the-art probabilistic risk assessment techniques, should continue to serve as the primary organizing principle of its regulatory framework. The Task Force concludes that the application of the defense-in-depth philosophy can be strengthened by including explicit requirements for beyond-design-basis events.

Many of the elements of such a regulatory framework already exist in the form of rules regarding station blackout, anticipated transient without scram, maintenance, combustible gas control, aircraft impact assessment, beyond-design-basis fires and explosions, and alternative treatment. Other elements, such as severe accident management guidelines, exist in voluntary industry initiatives. The Task Force has concluded that a collection of such "extended design-basis" requirements, with an appropriate set of quality or special treatment standards, should be established.

The Task Force further sees this approach, if implemented, as a more comprehensive and systematic application of defense-in-depth to NRC requirements for providing "adequate protection" of public health and safety. Implementation of this concept would require strong Commission support for a clear policy statement, rule changes, and revised staff guidance.

The Task Force notes that, after the attacks of September 11, 2001, the Commission established new security requirements on the basis of adequate protection. These new requirements did not result from any immediate or imminent threat to NRC-licensed facilities, but rather from new insights regarding potential security events. The Task Force concluded that the Fukushima Dai-ichi accident similarly provides new insights regarding low-likelihood, high-consequence events that warrant enhancements to defense-in-depth on the basis of redefining the level of protection that is regarded as adequate. The Task Force recommendation for an enhanced regulatory framework is intended to establish a coherent and transparent basis for treatment of the Fukushima insights. It is also intended to provide lasting direction to the staff regarding a consistent decisionmaking framework for future issues.

The Task Force has considered industry initiatives in this framework and sees that these could play a useful and valuable role. The Task Force believes that voluntary industry initiatives should not serve as a substitute for regulatory requirements but as a mechanism for facilitating and standardizing implementation of such requirements.

The Task Force applied this conceptual framework during its deliberations. The result is a set of recommendations that take a balanced approach to defense-in-depth as applied to low-likelihood, high-consequence events such as prolonged station blackout resulting from severe natural phenomena. These recommendations, taken together, are intended to clarify and strengthen the regulatory framework for protection against natural disasters, mitigation, and emergency preparedness, and to improve the effectiveness of the NRC's programs. The Task Force's overarching recommendations are:

Clarifying the Regulatory Framework

1. The Task Force recommends establishing a logical, systematic, and coherent regulatory framework for adequate protection that appropriately balances defense-in-depth and risk considerations. (Section 3)

Ensuring Protection

2. The Task Force recommends that the NRC require licensees to reevaluate and upgrade as necessary the design-basis seismic and flooding protection of structures, systems, and components for each operating reactor. (Section 4.1.1)

3. The Task Force recommends, as part of the longer term review, that the NRC evaluate potential enhancements to the capability to prevent or mitigate seismically induced fires and floods. (Section 4.1.2)

Enhancing Mitigation

4. The Task Force recommends that the NRC strengthen station blackout mitigation capability at all operating and new reactors for design-basis and beyond-design-basis external events. (Section 4.2.1)

5. The Task Force recommends requiring reliable hardened vent designs in boiling water reactor facilities with Mark I and Mark II containments. (Section 4.2.2)

6. The Task Force recommends, as part of the longer term review, that the NRC identify insights about hydrogen control and mitigation inside containment or in other buildings as additional information is revealed through further study of the Fukushima Dai-ichi accident. (Section 4.2.3)

7. The Task Force recommends enhancing spent fuel pool makeup capability and instrumentation for the spent fuel pool. (Section 4.2.4)

8. The Task Force recommends strengthening and integrating onsite emergency response capabilities such as emergency operating procedures, severe accident management guidelines, and extensive damage mitigation guidelines . (Section 4.2.5)

Strengthening Emergency Preparedness

9. The Task Force recommends that the NRC require that facility emergency plans address prolonged station blackout and multiunit events. (Section 4.3.1)

10. The Task Force recommends, as part of the longer term review, that the NRC pursue additional emergency preparedness topics related to multiunit events and prolonged station blackout. (Section 4.3.1)

11. The Task Force recommends, as part of the longer term review, that the NRC should pursue emergency preparedness topics related to decisionmaking, radiation monitoring, and public education. (Section 4.3.2)

Improving the Efficiency of NRC Programs

12. The Task Force recommends that the NRC strengthen regulatory oversight of licensee safety performance (i.e., the Reactor Oversight Process) by focusing more attention on defense-in-depth requirements consistent with the recommended defense-in-depth framework. (Section 5.1)

The Task Force presents further details on its recommendations in this report and an implementation strategy in Appendix A. The strategy includes several rulemaking activities to establish new requirements. Recognizing that rulemaking and subsequent implementation typically take several years to accomplish, the Task Force recommends interim actions to enhance protection, mitigation, and preparedness while the rulemaking activities are conducted.

These recommendations are based on the best available information regarding the Fukushima Dai-ichi accident and a review of relevant NRC requirements and programs. The Task Force concludes that these are a reasonable set of actions to enhance U.S. reactor safety in the 21st century.

1. INTRODUCTION

In the days following the Fukushima Dai-ichi nuclear accident in Japan, the U.S. Nuclear Regulatory Commission (NRC) directed the staff to establish a senior-level agency task force (the Task Force) to conduct a methodical and systematic review of the NRC's processes and regulations to determine whether the agency should make additional improvements to its regulatory system and to make recommendations to the Commission for its policy direction. The Commission direction was provided in a tasking memorandum dated March 23, 2011, from Chairman Gregory B. Jaczko to the Executive Director for Operations. The tasking included objectives for a near-term and a longer term review. Appendix B provides a copy of the tasking memorandum.

In response to the Commission's direction, the Executive Director for Operations established an agency Task Force to conduct a near-term evaluation of the need for agency actions. The Task Force's charter, dated March 30, 2011, called for a methodical and systematic review of relevant NRC regulatory requirements, programs, and processes, and their implementation, and to recommend whether the agency should make near-term improvements to its regulatory system. The charter also directed the Task Force to identify topics for review and assessment for the longer term effort. Appendix C presents a copy of the Task Force charter.

The Task Force structured its review activities to reflect insights from past lessons-learned efforts. For example, after the 1979 accident at Three Mile Island (TMI), the NRC had not yet developed much of the decisionmaking framework that is in place today. In addition, the post-TMI review considered a number of actions that were proposed for general safety enhancement rather than being directed at specific safety weaknesses revealed by the TMI accident. As a result, some of the actions taken by the NRC after TMI were not subjected to a structured review and were subsequently not found to be of substantial safety benefit and were removed.

In establishing a systematic and methodical process for its review, the Task Force determined that it would focus its efforts on areas that had a nexus to the Fukushima Dai-ichi accident regarding the initiating event, the response of equipment and personnel, and the progression of the accident as the Task Force understands it from available information. The Task Force recognized that detailed information in each of these areas was, in many cases, unavailable, unreliable, or ambiguous because of damage to equipment at the site and because the Japanese response continues to focus on actions to stop the ongoing radioactive release and to achieve long-term core and spent fuel pool cooling. Even without a detailed understanding of all aspects of the Fukushima accident, the Task Force identified those key areas most relevant to the safety of U.S. reactors, such as external events that could damage large areas of the plant, protection against and mitigation of a prolonged station blackout, and management of severe accidents. Section 2 of this report presents the sequence of events at Fukushima, as relevant to the Task Force's deliberations. The Task Force did not address insights from the NRC's incident response activities related to the Fukushima accident as these are being addressed by the NRC's Office of Nuclear Security and Incident Response.

Consistent with its tasking and charter, the Task Force remained independent of industry efforts, while obtaining a broad range of inputs. The Task Force had full access to the NRC staff to obtain information on existing programs, received briefings from staff experts in the Headquarters offices, and solicited inputs from all four NRC regional offices. The Task Force also obtained valuable insights from the members of the NRC site team in Japan. The Task Force requested information on the status of licensees' implementation of severe accident management guidelines (SAMGs); the regional offices collected this information through Temporary Instruction (TI) 2515/184, "Availability and Readiness Inspection of Severe Accident Management Guidelines (SAMGs)," dated April 29, 2011, and the Office of Nuclear Reactor Regulation compiled the information for the Task Force. During the implementation of this TI, members of the Task Force accompanied the inspectors at two nuclear power plant sites to gain an independent perspective and additional insights into licensee decisionmaking during severe accidents. While maintaining its independence of industry efforts, members of the Task Force met with representatives of the Institute of Nuclear Power Operations to gather information on the industry's post-Fukushima actions. The Task Force also met with representatives of the Federal Emergency Management Agency (FEMA) to discuss offsite emergency preparedness and to obtain insights on the U.S. National Response Framework. Finally, the Task Force appropriately screened and considered information and suggestions received from internal and external stakeholders. The Task Force monitored, directly or indirectly, related international activities of the International Atomic Energy Agency (IAEA), Nuclear Energy Agency (NEA), and other organizations.

In developing recommendations, the Task Force considered the existing approach for regulatory decisionmaking. This approach includes the existing technical requirements related to the licensing, operation, and maintenance of commercial nuclear power plants. It is informed by the Commission's Policy Statement on Safety Goals for the Operations of Nuclear Power Plants, which appeared in the *Federal Register* in August 1986 (51 FR 30028). The approach includes the agency's historical commitment to a defense-in-depth philosophy that ensures that the design basis includes multiple layers of defense. In developing its recommendations, the Task Force was also guided by the NRC's Principles of Good Regulation, particularly as related to ensuring a clear nexus between the Commission's requirements and its goals and objectives. In addition, the Task Force was appropriately informed by the "Regulatory Analysis Guidelines of the U.S. Nuclear Regulatory Commission" (NUREG/BR-0058, Revision 4, issued September 2004), which provide a framework for evaluating potential new requirements.

The Policy Statement on Safety Goals sets forth two qualitative safety goals, which are supported by two quantitative supporting objectives. The following are the qualitative safety goals:

> Individual members of the public should be provided a level of protection from the consequences of nuclear power plant operation such that individuals bear no significant additional risk to life and health.

> Societal risks to life and health from nuclear power plant operation should be comparable to or less than the risks of generating electricity by viable competing technologies and should not be a significant addition to other societal risks.

The quantitative supporting objectives are as follows:

> The risk to an average individual in the vicinity of a nuclear power plant of prompt fatalities that might result from reactor accidents should not exceed one-tenth of one percent (0.1 percent) of the sum of prompt fatality risks resulting from other accidents to which members of the U.S. population are generally exposed.

The risk to the population in the area near a nuclear power plant of cancer fatalities that might result from nuclear power plant operation should not exceed one-tenth of one percent (0.1 percent) of the sum of cancer fatality risks resulting from all other causes.

In the Policy Statement on Safety Goals, the Commission emphasized the importance of features such as containment, siting, and emergency planning as "integral parts of the defense-in-depth concept associated with its accident prevention and mitigation philosophy." A cursory review of documents discussing the agency's approach to defense-in-depth provides a range of explanations and applications.

The Commission's policy on probabilistic risk assessment (PRA) ("Use of Probabilistic Risk Assessment Methods in Nuclear Regulatory Activities," dated August 16, 1995), states the following:

Defense-in-depth is a philosophy used by the NRC to provide redundancy for facilities with "active" safety systems, e.g. a commercial nuclear power [plant], as well as the philosophy of a multiple-barrier approach against fission product releases.

An instructive discussion of the defense-in-depth philosophy also appears in director's decisions relating to a petition on Davis-Besse (FirstEnergy Nuclear Operating Company (Davis-Besse Nuclear Power Station, Unit 1), DD-03-3, 58 NRC 151, 163 (2003)).

The decision described defense-in-depth as encompassing the following requirements:

(1) require the application of conservative codes and standards to establish substantial safety margins in the design of nuclear plants;

(2) require high quality in the design, construction, and operation of nuclear plants to reduce the likelihood of malfunctions, and promote the use of automatic safety system actuation features;

(3) recognize that equipment can fail and operators can make mistakes and, therefore, require redundancy in safety systems and components to reduce the chance that malfunctions or mistakes will lead to accidents that release fission products from the fuel;

(4) recognize that, in spite of these precautions, serious fuel-damage accidents may not be completely prevented and, therefore, require containment structures and safety features to prevent the release of fission products; and

(5) further require that comprehensive emergency plans be prepared and periodically exercised to ensure that actions can and will be taken to notify and protect citizens in the vicinity of a nuclear facility.

The Task Force has found that the defense-in-depth philosophy is a useful and broadly applied concept. It is not, however, susceptible to a rigid definition because it is a philosophy. For the purposes of its review, the Task Force focused on the following application of the defense-in-depth concept:

- protection from external events that could lead to fuel damage
- mitigation of the consequences of such accidents should they occur, with a focus on preventing core and spent fuel damage and uncontrolled releases of radioactive material to the environment

- emergency preparedness (EP) to mitigate the effects of radiological releases to the public and the environment, should they occur

These levels of defense-in-depth are appropriate for significant external challenges to a facility. In applying these defense-in-depth features, the Task Force sought to ensure that the Commission's regulatory requirements, processes, and programs effectively address each layer of protection while maintaining appropriate balance among them. The Task Force notes that this approach is also consistent with Levels of Defense 3, 4, and 5 in IAEA Draft Safety Standard DS 414, "Safety of Nuclear Power Plants: Design," dated January 2010.

The framework of the Regulatory Analysis Guidelines (NUREG/BR-0058) informed the Task Force as it developed recommendations and evaluated potential new requirements. These guidelines provide a range of potential actions depending on the nature of the underlying safety issue and the means of addressing it. These potential actions include the following:

- actions necessary to bring a facility into compliance with existing requirements (Title 10 of the *Code of Federal Regulations* (10 CFR) 50.109(a)(4)(i))

- actions necessary to ensure adequate protection (10 CFR 50.109(a)(4)(ii))

- actions defining or redefining what level of protection of the public health and safety or common defense and security should be regarded as adequate (10 CFR 50.109(a)(4)(iii))

- actions that provide substantial additional protection and for which the direct and indirect costs are justified by the increased protection afforded (10 CFR 50.109(a)(3))

In developing its recommendations, the Task Force considered all of the above-mentioned guidance (defense-in-depth, the Policy Statement on Safety Goals, Regulatory Analysis Guidelines, and the backfit rule as codified in 10 CFR 50.109, "Backfitting"). In the Task Force's deliberations, it became apparent that the existing guidance does not present a completely clear and consistent framework for decisionmaking. The Policy Statement on Safety Goals and Regulatory Analysis Guidelines address only limited aspects of defense-in-depth. The safety goals policy addresses defense-in-depth by setting a "subsidiary numerical objective" (a third-level goal) for core damage frequency, in addition to the higher level health effects goals. In the Regulatory Analysis Guidelines, which provide guidance on implementation of the backfit rule, the phrase "defense-in-depth" occurs only twice in 46 pages of guidance; those references are limited to dealing with core damage frequency and containment performance in the screening criteria. The defense-in-depth concept is notably absent from the Regulatory Analysis Guidelines as part of the decision rationale.

Extensive efforts have been made in the past to address safety goals and their relationship to defense-in-depth. The most extensive attempt occurred as part of the technology-neutral framework effort described in NUREG-1860, "Feasibility Study for a Risk-Informed and Performance-Based Regulatory Structure for Future Plant Licensing," issued December 2007. The technology-neutral framework effort attempted to reconcile these issues in a single process in which defense-in-depth plays a fundamental role in addressing and compensating for uncertainties.

Ultimately, the Task Force chose a decision rationale built around the defense-in-depth concept in which each level of defense-in-depth (namely protection, mitigation, and EP) is critically evaluated for its completeness and effectiveness in performing its safety function. The Task Force has therefore developed a comprehensive set of recommendations to increase safety and redefine what level of protection of the public health is regarded as adequate. The Task Force's logic is that (1) these recommendations are intended to make each level of defense-in-depth complete and effective and (2) ensuring the completeness and effectiveness of each level of defense-in-depth is an essential element in the overall approach to ensuring safety, called "adequate protection." With this in mind, the Task Force has recommended what it concludes are significant reinforcements to NRC requirements and programs.

Finally, the Task Force considered its recommendations and the Commission's requirements for nuclear power plants in the context of the NRC's Principles of Good Regulation. In particular, the principle of "clarity" states: "Regulations should be coherent, logical, and practical. There should be a clear nexus between regulations and agency goals and objectives, whether explicitly or implicitly stated. Agency positions should be readily understood and easily applied." Also, the principle of "efficiency" states, in part, that "Regulatory activities should be consistent with the degree of risk reduction they achieve."

The principles of "independence" and "openness" focus on the importance of obtaining inputs from the full range of stakeholders, including consideration of many and possibly conflicting public interests, and open channels of communication. The duration and scope of the Task Force's effort have necessarily limited the degree of stakeholder interaction that was possible. The implementation of Task Force recommendations will require additional effort by the NRC staff to conduct stakeholder outreach through its normal processes (e.g., rulemaking, licensing, public meetings, and workshops).

This page intentionally left blank

2.   SUMMARY OF EVENTS AT FUKUSHIMA DAI-ICHI

At 14:46 Japan standard time on March 11, 2011, the Great East Japan Earthquake—rated a magnitude 9.0—occurred at a depth of approximately 25 kilometers (15 miles), 130 kilometers (81 miles) east of Sendai and 372 kilometers (231 miles) northeast of Tokyo off the coast of Honshu Island. This earthquake resulted in the automatic shutdown of 11 nuclear power plants at four sites along the northeast coast of Japan (Onagawa 1, 2, and 3; Fukushima Dai-ichi 1, 2, and 3; Fukushima Dai-ni 1, 2, 3, and 4; and Tokai 2). The earthquake precipitated a large tsunami that is estimated to have exceeded 14 meters (45 feet) in height at the Fukushima Dai-ichi Nuclear Power Plant site. The earthquake and tsunami produced widespread devastation across northeastern Japan, resulting in approximately 25,000 people dead or missing, displacing many tens of thousands of people, and significantly impacting the infrastructure and industry in the northeastern coastal areas of Japan.

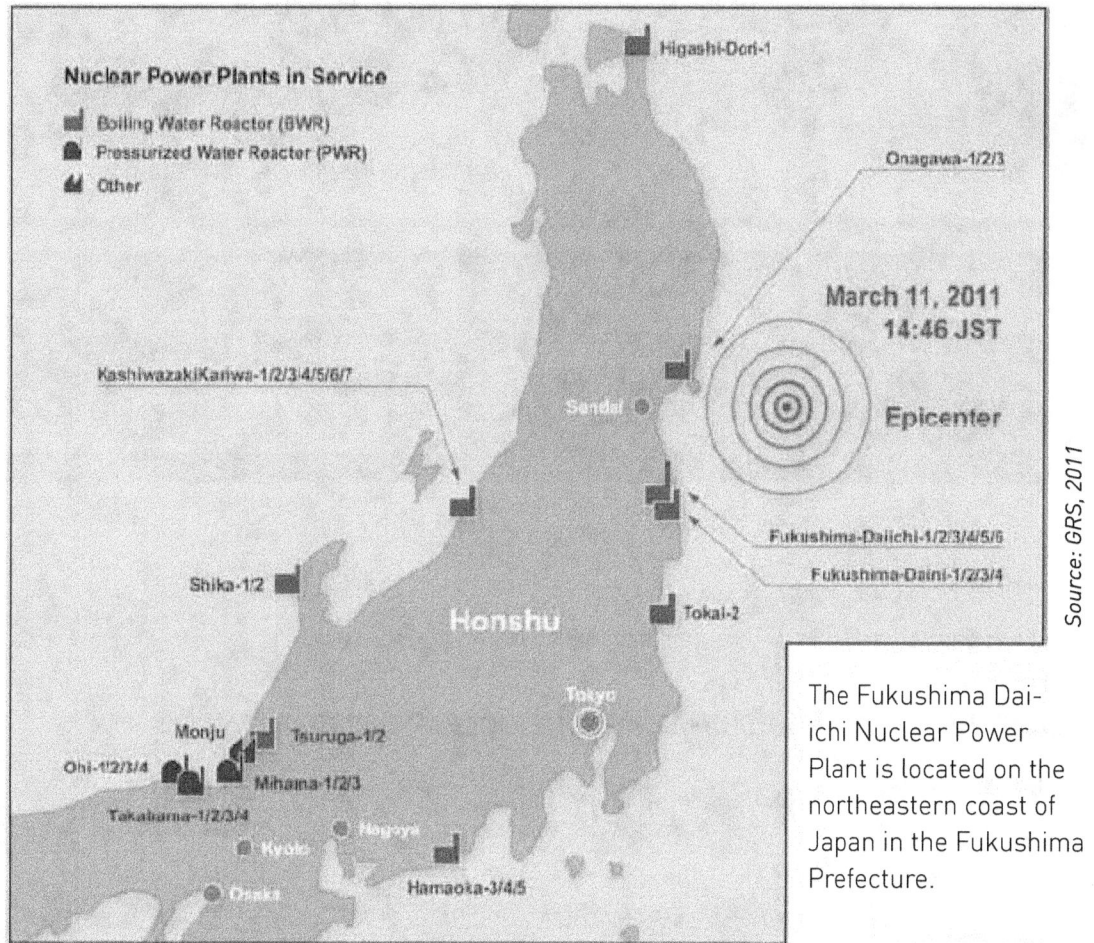

The Fukushima Dai-ichi Nuclear Power Plant is located on the northeastern coast of Japan in the Fukushima Prefecture.

Source: GRS, 2011

Figure 1: Nuclear Power Plants in Japan

Fukushima Dai-ichi Units 1 through 4 are located in the southern part of the station and are oriented such that Unit 1 is the northernmost and Unit 4 is the southernmost. Fukushima Dai-ichi Units 5 and 6 are located farther north and at a somewhat higher elevation than the Unit 1–4 cluster, and Unit 6 is located to the north of Unit 5.

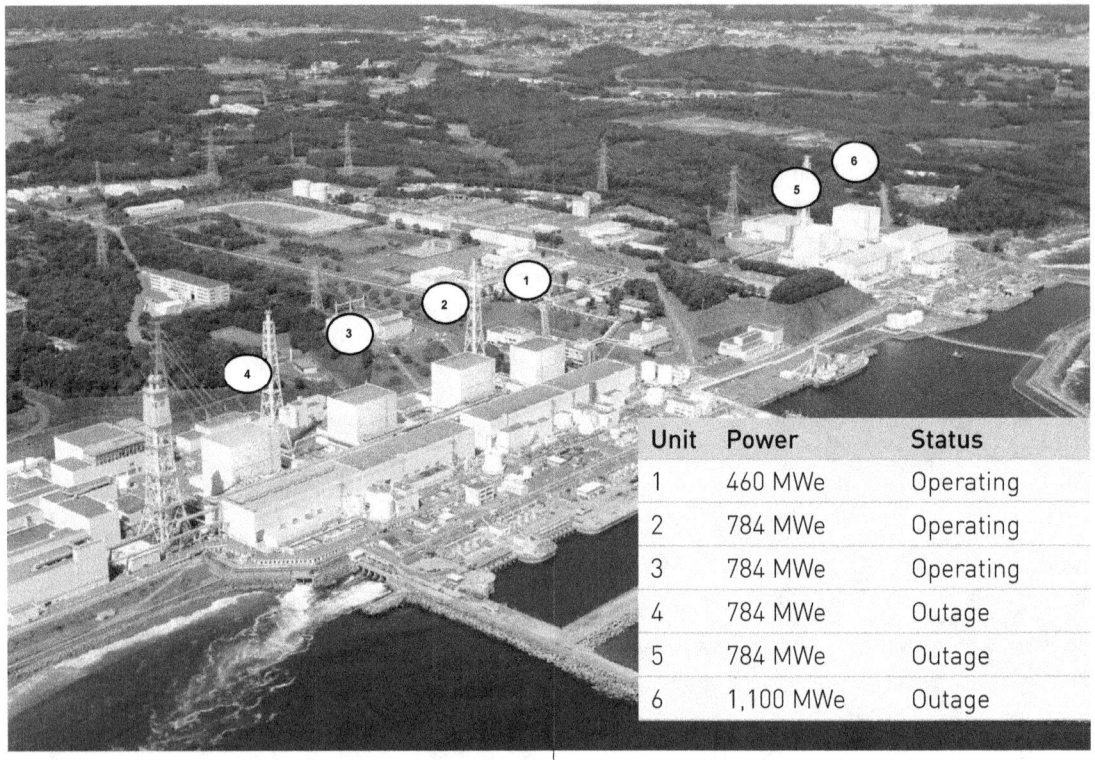

Unit	Power	Status
1	460 MWe	Operating
2	784 MWe	Operating
3	784 MWe	Operating
4	784 MWe	Outage
5	784 MWe	Outage
6	1,100 MWe	Outage

Figure 2: Fukushima Dai-ichi Nuclear Power Plant (status before earthquake)

The Fukushima Dai-ichi site includes six boiling water reactors (BWRs).

REACTORS AT THE FUKUSHIMA DAI-ICHI NUCLEAR POWER PLANT

Unit	Net MWe*	Reactor, Containment, and Cooling Systems**
1	460	BWR-3, Mark I, IC, HPCI
2	784	BWR-4, Mark I, RCIC, HPCI
3	784	BWR-4, Mark I, RCIC, HPCI
4	784	BWR-4, Mark I, RCIC, HPCI
5	784	BWR-4, Mark I, RCIC, HPCI
6	1,100	BWR-5, Mark II, RCIC, HPCS

* *MWe—megawatts electric*

** *IC—isolation condenser, HPCI—high-pressure coolant injection system, RCIC—reactor core isolation cooling system, HPCS—high-pressure core spray system*

On March 11, 2011, Units 1, 2, and 3 were in operation, and Units 4, 5, and 6, were shut down for routine refueling and maintenance activities; the Unit 4 reactor fuel was offloaded to the Unit 4 spent fuel pool.

As a result of the earthquake, all of the operating units appeared to experience a normal reactor trip within the capability of the safety design of the plants. The three operating units at Fukushima Dai-ichi automatically shut down, apparently inserting all control rods into the reactor. As a result of the earthquake, offsite power was lost to the entire facility. The emergency diesel generators started at all six units providing alternating current (ac) electrical power to critical systems at each unit, and the facility response to the seismic event appears to have been normal.

Approximately 40 minutes following the earthquake and shutdown of the operating units, the first large tsunami wave inundated the site followed by multiple additional waves. The estimated height of the tsunami exceeded the site design protection from tsunamis by approximately 8 meters (27 feet). The tsunami resulted in extensive damage to site facilities and a complete loss of ac electrical power at Units 1 through 5, a condition known as station blackout (SBO). Unit 6 retained the function of one of the diesel generators.

The operators were faced with a catastrophic, unprecedented emergency situation. They had to work in nearly total darkness with very limited instrumentation and control systems. The operators were able to successfully cross-tie the single operating Unit 6 air-cooled diesel generator to provide sufficient ac electrical power for Units 5 and 6 to place and maintain those units in a safe shutdown condition, eventually achieving and maintaining cold shutdown.

Despite the actions of the operators following the earthquake and tsunami, cooling was lost to the fuel in the Unit 1 reactor after several hours, the Unit 2 reactor after about 71 hours, and the Unit 3 reactor after about 36 hours, resulting in damage to the nuclear fuel shortly after the loss of cooling. Without ac power, the plants were likely relying on batteries and turbine-driven and diesel-driven pumps. The operators were likely implementing their severe accident management program to maintain core cooling functions well beyond the normal capacity of the station batteries. Without the response of offsite assistance, which appears to have been hampered by the devastation in the area, among other factors, each unit eventually lost the capability to further extend cooling of the reactor cores.

The current condition of the Unit 1, 2, and 3 reactors is relatively static, but those units have yet to achieve a stable, cold shutdown condition. Units 1, 2, 3, and 4 also experienced explosions further damaging the facilities and primary and secondary containment structures. The Unit 1, 2, and 3 explosions were caused by the buildup of hydrogen gas within primary containment produced during fuel damage in the reactor and subsequent movement of that hydrogen gas from the drywell into the secondary containment. The source of the explosive gases causing the Unit 4 explosion remains unclear. In addition, the operators were unable to monitor the condition of and restore normal cooling flow to the Unit 1, 2, 3, and 4 spent fuel pools.

Below is a sequence of events early in the accident for the six Fukushima Dai-ichi reactors. Only information from the Japanese utility and official Japanese Government sources, including the "Report of the Japanese Government to the IAEA Ministerial Conference on Nuclear Safety," is included. More detail is expected to emerge as recovery from the accident continues and access to various locations, facilities, data, and staff improves. This sequence of events provides only the level of detail necessary for the near-term assessment of insights and the recommendation of actions for consideration at U.S. nuclear facilities. When available, times indicated are in Japan standard time.

Unit 1 Sequence of Events

March 11

14:47	Earthquake, loss of offsite ac power, and plant trip
14:52	Isolation condenser operated to cool reactor
15:03	Isolation condenser stopped operating
15:37	Tsunami and total loss of ac power—SBO
15:37	Loss of ability to inject water to the reactor
~17:00	Water level below top of fuel
--:--	Partial core damage (several hours after tsunami)

March 12

14:30	Vent primary containment
15:36	Explosion results in severe damage to the reactor building (secondary containment)

Figure 3: Damage to Unit 1

Enhancing Reactor Safety in the 21st Century

Unit 2 Sequence of Events

March 11

14:47	Earthquake, loss of offsite ac power, and plant trip
~14:50	RCIC manually operated to inject water to reactor
15:41	Tsunami and total loss of ac power at site—SBO

March 13

--:--	RCIC continued to be used to cool reactor
~11:00	Vent primary containment

March 14

13:25	RCIC stopped operating
~18:00	Water level below top of fuel
--:--	Partial core damage (approximately 3 days after tsunami)
--:--	Blowout panel open on side of reactor building

March 15

~06:00	Explosion; suppression chamber pressure decreased indicating the possibility that primary containment was damaged

Figure 4: Damage to Unit 2

Unit 3 Sequence of Events

March 11

14:47	Earthquake, loss of offsite ac power, and plant trip
15:05	RCIC manually started to inject water into reactor
15:41	Tsunami and total loss of ac power at site—SBO

March 12

11:36	RCIC stopped operating
12:35	HPCI automatically started injecting water into reactor

March 13

02:42	HPCI stopped operating
~08:00	Water level below top of fuel
--:--	Partial core damage (approximately 2 days after tsunami)

March 14

05:20	Vent primary containment
11:01	Explosion results in severe damage to the reactor building (secondary containment)

Figure 5: Damage to Unit 3

Enhancing Reactor Safety in the 21st Century

Unit 4 Sequence of Events (Unit 4 reactor was defueled)

March 11		
	14:46	Earthquake and loss of offsite ac power
	15:38	Tsunami and total loss of ac power at site—SBO
March 15		
	~06:00	Explosion in reactor building

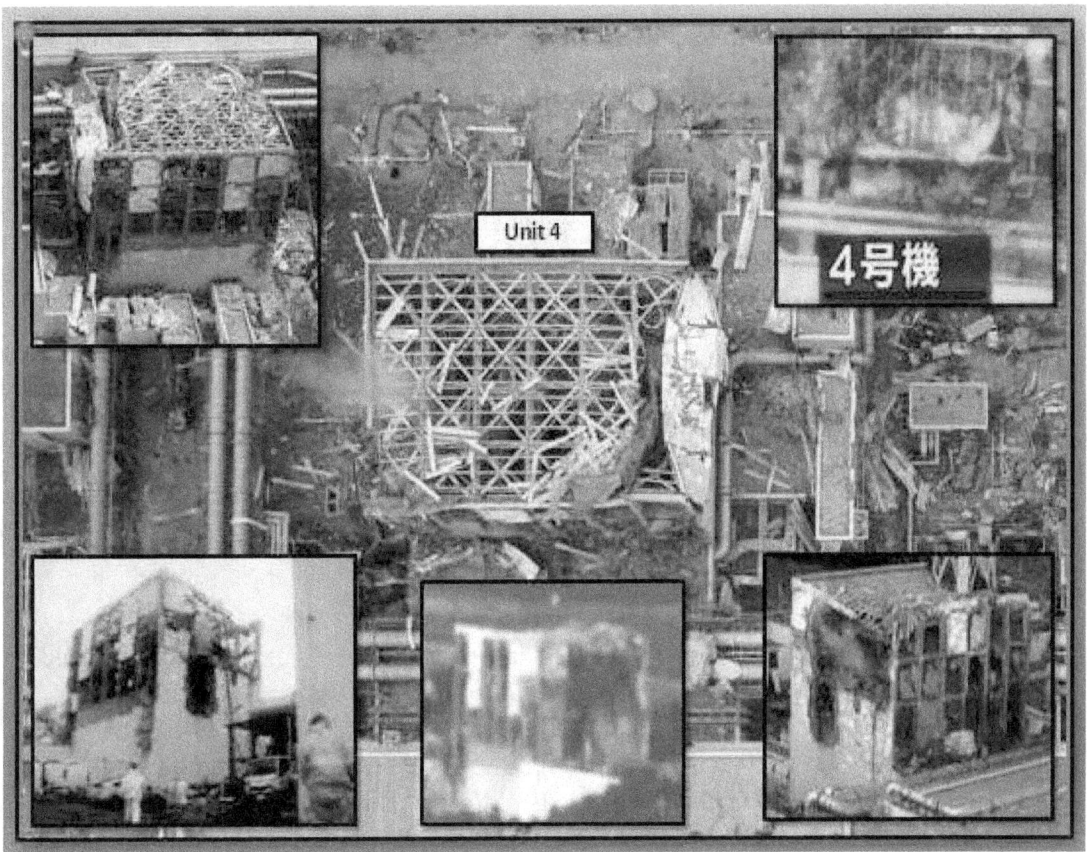

Figure 6: Damage to Unit 4

Unit 5 & 6 Sequence of Events (Both units were shut down for periodic inspection)

March 11

	14:46	Earthquake and loss of offsite ac power
	15:41	Tsunami and total loss of ac power at site—SBO

March 20

	14:30	Unit 5 enters cold shutdown
	19:27	Unit 6 enters cold shutdown

Protective Action Recommendations at Fukushima Dai-ichi

March 11

Evacuation of residents within 3 kilometers (1.9 miles) and shelter-in-place for residents within 10 kilometers (6.2 miles)

March 12

Evacuation of residents within 20 kilometers (12.4 miles)

March 15

Evacuation of residents within 30 kilometers (18.6 miles)

April 11

"Planned Evacuation Areas" and "Evacuation Prepared Area" established in the areas beyond 20 kilometers (12.4 miles)

April 21

Restricted area within 20 kilometers (12.4 miles) established to allow temporary access and exclusion area of 3 kilometers (1.9 miles) established for members of the public

As directed in the Chairman's tasking memorandum, the Task Force emphasizes the need for the staff to maintain awareness and develop further insights during the long-term followup of the Fukushima event. Additional information is expected to evolve and become better understood regarding (1) the condition of structures, systems, and components (SSCs) in Units 1 through 6, including the effects of the seismic and flooding events and the explosions, (2) the timing and success of the actions taken by facility staff, including the adequacy of procedures and SAMGs, and (3) the progression of the accidents in the Unit 1, 2, and 3 reactors and the Unit 1, 2, 3, and 4 spent fuel pools.

Figure 7: Fukushima Dai-ichi Units 1–4 following explosions

3. REGULATORY FRAMEWORK FOR THE 21ST CENTURY

BACKGROUND

The combination of the massive earthquake and devastating tsunami at Fukushima were well in excess of external events considered in the plant design. The Fukushima accident also challenged the plant's mitigation capabilities and EP.

With that in mind, this section addresses the elements of the NRC regulatory framework that play a part in providing protection from design-basis events, as well as events as severe and complex as the Fukushima accident. Those elements include protection against seismic and flooding events (characterized as design-basis events), protection for loss of all ac power (characterized as a beyond-design-basis event), and mitigation of severe accidents (addressing beyond-design-basis topics of core damage and subsequent containment performance), as well as EP. The Fukushima accident therefore highlights the full spectrum of considerations necessary for a comprehensive and coherent regulatory framework.

Similar issues were raised by the TMI accident, and many beyond-design-basis requirements, programs, and practices were derived from that experience and from the concurrent development of PRA as a practical tool. Other sections of this report address specific elements of protection, mitigation, and preparedness and evaluate their current capabilities, limitations, and potential enhancements. This section will evaluate the overall NRC regulatory approach for ensuring safety through a defense-in-depth philosophy that includes design-basis requirements and additional risk reduction requirements and programs. The goal of this section is to identify possible ways to better integrate the fundamental design-basis concept with the beyond-design-basis considerations, including EP.

Design-basis events became a central element of the safety approach almost 50 years ago when the U.S. Atomic Energy Commission (AEC) formulated the idea of requiring safety systems to address a prescribed set of anticipated operational occurrences and postulated accidents. In addition, the design-basis requirements for nuclear power plants included a set of external challenges including seismic activity and flooding from various sources. That approach and its related concepts of design-basis events and design bases were used in licensing the current generation of nuclear plants in the 1960s and 1970s.

Frequently, the concept of design-basis events has been equated to adequate protection, and the concept of beyond-design-basis events has been equated to beyond adequate protection (i.e., safety enhancements). This vision of adequate protection has typically only led to requirements addressing beyond-design-basis concerns when they were found to be associated with a substantial enhancement in safety and justified in terms of cost.

Starting in the 1980s and continuing to the present, the NRC has maintained the design-basis approach and expanded it to address issues of concern. The NRC added requirements to address each new issue as it arose but did not change the fundamental concept of design-basis events or the list of those events; nor did the NRC typically assign the concept of adequate protection to these changes. The following paragraphs include discussion of the historical development of requirements to address issues beyond the design basis, including the potential loss of all ac power (i.e., SBO) and other issues. In reading these paragraphs, it is helpful to keep in mind that actions to address issues beyond the design basis were largely considered safety enhancements. They were considered to be beyond what was required for adequate protection.

In the early 1970s, the AEC began questioning applicants and licensees about the potential consequences of an anticipated operational occurrence (a design-basis event) with the beyond-design-basis failure of the shutdown system (a sequence called anticipated transient without scram (ATWS)). This issue was clearly outside the design basis of the plants, yet it remained a concern for many years, eventually resulting in a regulation (10 CFR 50.62, "Requirements for Reducing the Risk from Anticipated Transients without Scram (ATWS) Events for Light-Water-Cooled Nuclear Power Plants"). That regulation established several requirements to address ATWS concerns but did not expand the list of design-basis events to include ATWS.

In the mid-1980s, the Commission established the SBO rule (10 CFR 50.63, "Loss of All Alternating Current Power") to address concerns related to loss of all ac power. Again the rule established new requirements but did not alter the design-basis concept or list of design-basis events. SBO is not a design-basis event, nor is SBO-related equipment subject to the NRC's quality assurance requirements in Appendix B, "Quality Assurance Criteria for Nuclear Power Plants and Fuel Reprocessing Plants," to 10 CFR Part 50, "Domestic Licensing of Production and Utilization Facilities."

In addition to the established regulatory approach, the AEC commissioned and the NRC later published the first PRA of nuclear power plants in WASH-1400, "Reactor Safety Study: An Assessment of Accident Risks in U.S. Commercial Nuclear Power Plants," issued October 1975. That study addressed possible reactor accidents based on the frequency of initiating events and their likely consequences. The study proceeded without regard to classification of events as "anticipated operational occurrences" or "postulated accidents" and therefore without regard to event categorization as "design-basis events." The treatment of events based solely on frequency and consequences remains a characteristic of PRA and is one of its strengths in complementing the traditional regulatory approach.

Following the TMI accident, numerous lessons-learned efforts were commissioned. One of those studies (NUREG/CR-1250, "Three Mile Island; A Report to the Commissioners and to the Public," issued in 1980 and generally referred to as the Rogovin Report) evaluated the then-existing NRC regulatory approach (characterized in the report as "the so-called 'design basis accident' concept"). The report concluded that "More rigorous and quantitative methods of risk analysis have been developed and should be employed to assess the safety of design and operation," and "The best way to improve the existing design review process is by relying in a major way upon quantitative risk analysis." The report's recommendations to the Commission included "Expand the spectrum of design basis accidents..." and "On a selective basis, determine whether some design features to mitigate the effects of some Class Nine accidents [i.e., severe accidents in current terminology] should be required."

On the basis of all the recommendations developed by the NRC following the TMI accident, the Commission established numerous new requirements (approximately 120 actions per plant), some within and some beyond the design bases. In addition, the Commission considered action to address core damage scenarios, again beyond the design basis. Ultimately, the Commission encouraged licensees to use the newly developed PRA methodology to search for vulnerabilities (in the Individual Plant Examination (IPE) program and Individual Plant Examination for External Events (IPEEE) program) and requested information on their findings. The Commission also encouraged the development of SAMGs based on PRA insights and severe accident research. However, the Commission did not take action to require the IPEs, IPEEEs, or SAMGs. While the Commission has been partially responsive to recommendations calling for requirements to address beyond-design-basis accidents, the NRC has not made fundamental changes to the regulatory approach for beyond-design-basis events and severe accidents for operating reactors.

Currently, risk-informed regulation (i.e., regulation using PRAs) serves the limited roles of maintenance rule implementation, Regulatory Analysis Guidelines, the search for vulnerabilities (e.g., through the IPE and IPEEE programs), the Reactor Oversight Process (ROP) and its significance determination process, and voluntary license amendment applications (e.g., risk-informed inservice inspection).

In contrast, for new reactors, the Commission has moved further from a largely design-basis accident concept, requiring applicants for design certifications and combined licenses (COLs) under 10 CFR Part 52, "Licenses, Certification, and Approvals for Nuclear Power Plants," to perform a PRA and provide a description and analysis of design features for the prevention and mitigation of severe accidents (10 CFR 52.47(23) and 10 CFR 52.79(48)). Each design certification rule (10 CFR Part 52, Appendix A, "Design Certification Rule for the U.S. Advanced Boiling Water Reactor," and other Part 52 appendices) then codifies the severe accident features of each approved standard design.

Following the terrorist events of September 11, 2001, the NRC issued security advisories, orders, license conditions, and ultimately a new regulation (10 CFR 50.54(hh)) to require licensees to develop and implement guidance and strategies to maintain or restore capabilities for core cooling and containment and spent fuel pool cooling under the circumstances associated with the loss of large areas of the plant due to a fire or explosion. These requirements have led to the development of extensive damage mitigation guidelines (EDMGs) at all U.S. nuclear power plants. The NRC has inspected the guidelines and strategies that licensees have implemented to meet the requirements of 10 CFR 50.54(hh)(2). However, there are no specific quality requirements associated with these requirements, and the quality assurance requirements of 10 CFR Part 50, Appendix B, do not apply. The EDMGs are requirements for addressing events well beyond those historically considered to be the design basis and were implemented as adequate protection backfits. In order to address the changing security threat environment, the Commission effectively redefined what level of protection should be regarded as adequate. This is a normal and reasonable, albeit infrequent, exercise of the NRC's responsibilities of protecting public health and safety.

All of the above indicate the Commission's desire and commitment to act either through regulatory requirements or voluntary industry initiatives to address concerns related to the design basis or beyond the design basis where appropriate.

TASK FORCE EVALUATION

As presented above, the current NRC regulatory approach includes (1) requirements for design-basis events with features controlled through specific regulations or the general design criteria (GDC) (10 CFR Part 50, Appendix A, "General Design Criteria for Nuclear Power Plants") and the quality requirements of 10 CFR Part 50, Appendix B, (2) beyond-design-basis requirements through specific rules (e.g., the SBO rule) with specified quality requirements, (3) voluntary industry initiatives to address severe accident features, strategies, and guidelines for operating reactors, and (4) specific requirements to address damage from fires and explosions and their mitigation.

The Task Force presents the following observations on the NRC regulatory approach:

- Although complex, the current regulatory approach has served the Commission and the public well and allows the Task Force to conclude that a sequence of events like those occurring in the Fukushima accident is unlikely to occur in the United States and could be mitigated, reducing the likelihood of core damage and radiological releases.

- Therefore, in light of the low likelihood of an event beyond the design basis of a U.S. nuclear power plant and the current mitigation capabilities at those facilities, the Task Force concludes that continued operation and continued licensing activities do not pose an imminent risk to the public health and safety and are not inimical to the common defense and security. Nonetheless, the Task Force is recommending building on the safety foundation laid in the 1960s and 1970s, and the safety improvements added from the 1980s to the present, to produce a regulatory structure well suited to licensing and overseeing the operation of nuclear power plants for decades to come. The Task Force sees these recommendations, not as a rejection of the past, but more as a fulfillment of past intentions.

- Adequate protection has been, and should continue to be, an evolving safety standard supported by new scientific information, technologies, methods, and operating experience. This was the case when new information about the security environment was revealed through the events of September 11, 2001. Licensing or operating a nuclear power plant with no emergency core cooling system or without robust security protections, while done in the past, would not occur under the current regulations. As new information and new analytical techniques are developed, safety standards need to be reviewed, evaluated, and changed, as necessary, to insure that they continue to address the NRC's requirements to provide reasonable assurance of adequate protection of public health and safety. The Task Force believes, based on its review of the information currently available from Japan and the current regulations, that the time has come for such change.

- In response to the Fukushima accident and the insights it brings to light, the Task Force is recommending actions, some general, some specific, that it believes would be a reasonable, well-formulated set of actions to increase the level of safety associated with adequate protection of the public health and safety.

- The Commission has come to rely on design-basis requirements and a patchwork of beyond-design-basis requirements and voluntary initiatives for maintaining safety. Design-basis requirements include consideration of anticipated operational occurrences and postulated accidents such as loss-of-coolant accidents. Beyond-design-basis considerations such as ATWS and SBO are discussed below. Voluntary initiatives have addressed some severe accident considerations (through the IPE and IPEEE programs), shutdown risk issues, containment vents for BWR Mark I designs, and SAMGs.

- The concept of beyond-design-basis requirements applies, for example, to ATWS, SBO, aircraft impact assessment (AIA), combustible gas control, and EDMGs. Since fire protection is not based on a design-basis fire, it too can be considered beyond design basis. Although the phrase "beyond design basis" appears only once in the NRC regulations (i.e., in 10 CFR 50.150, "Aircraft Impact Assessment," known as the AIA rule), regulators and industry use it often. Unfortunately, the phrase "beyond design basis" is vague, sometimes misused, and often misunderstood. Several elements of the phrase contribute to these misunderstandings. First, some beyond-design-basis considerations have been incorporated into the requirements and therefore directly affect reactor designs. The phrase is therefore inconsistent with the normal meaning of the words. In addition, there are many other beyond-design-basis considerations that are not requirements. The phrase therefore fails to convey the importance of the requirements to which it refers.

- The Task Force has noted that other international regulatory systems also address considerations beyond the design basis. For example, while the NRC addresses regulatory requirements in five categories—three design basis (normal operation, anticipated operational occurrences, and postulated accidents) and two beyond design basis (one required and one voluntary)—Finland addresses regulatory requirements in six categories—four design basis (normal operation, anticipated operational occurrences, and two postulated accident categories), one "design extension condition," plus severe accidents. France also addresses both design-basis requirements and additional requirements in categories called "Risk Reduction Category 1" and "Risk Reduction Category 2." In addition, the phrase used in the IAEA Draft Safety Standard DS 414 addresses considerations beyond the design basis, referring to them as those addressing "design extension conditions." In this report, the Task Force will refer to past considerations beyond the design basis using that phrase (e.g., "beyond-design-basis events"). In the context of the Task Force recommendation for a new regulatory framework for the future, the Task Force will refer to such considerations as "extended design basis" requirements.

- The primary responsibility for safety rests with licensees, and the NRC holds licensees accountable for meeting regulatory requirements. In addition, voluntary safety initiatives by licensees can enhance safety if implemented and maintained effectively, but should not take the place of needed regulatory requirements. The NRC inspection and licensing programs give less attention to beyond-design-basis requirements and little attention to industry voluntary initiatives since there are no requirements to inspect against. Because of this, the NRC gives much more attention to design-basis events than to severe accidents.

- With the exception of a few special cases, licensees of operating reactors are not required to develop or maintain a PRA, although all licensees currently have a PRA. These PRAs are of varying scope and are generally not required to meet NRC-endorsed quality standards. New reactor applications must include a description of a design-specific PRA and its results and must address severe accident protection and mitigation features.

- The Commission has expressed its intent with respect to industry initiatives in the Regulatory Analysis Guidelines (NUREG/BR-0058, Revision 4). That document states, "It must be clear to the public that substituting industry initiatives for NRC regulatory action can provide effective and efficient resolution of issues, will in no way compromise plant safety, and does not represent a reduction in the NRC's commitment to safety and sound regulation."

- Lastly, the Task Force observes that for new reactor designs, the Commission's expectations that beyond-design-basis and severe accident concerns be addressed and resolved at the design stage are largely expressed in policy statements and staff requirements memoranda, only reaching the level of rulemaking when each design is codified through design certification rulemaking.

In summary, the major elements of the NRC regulatory approach relevant to the Fukushima accident, or a similar accident in the United States, are seismic and flooding protection (well established in the design-basis requirements); SBO protection (required, but beyond the design-basis requirements); and severe accident mitigation (expected but neither the severe accident mitigation features nor the SAMGs are required). In addition, U.S. facilities could employ EDMGs as further mitigation capability. The Task Force observes that this collection of approaches is largely the product of history; it was developed for the purpose of reactor licensing in the 1960s and 1970s and supplemented as necessary to address significant events or new issues. This evolution has resulted in a patchwork regulatory approach.

The Fukushima accident clearly demonstrates the importance of defense-in-depth. Whether through extraordinary circumstances or through limited knowledge of the possibilities, plants can be challenged beyond their established design bases protection. In such circumstances, the next layer of defense-in-depth, mitigation, is an essential element of adequate protection of public health and safety. Mitigation is provided for beyond-design-basis events and severe accidents, both of which involve external challenges or multiple failures beyond the design basis. This beyond-design-basis layer of defense-in-depth is broadly consistent with the IAEA concept of "design extension conditions" (presented in Draft Safety Standard DS 414).

The Task Force concludes that the NRC's safety approach is incomplete without a strong program for dealing with the unexpected, including severe accidents. Continued reliance on industry initiatives for a fundamental level of defense-in-depth similarly would leave gaps in the NRC regulatory approach. The Commission has clearly established such defense-in-depth severe accident requirements for new reactors (in 10 CFR 52.47(23), 10 CFR 52.79(38), and each design certification rule), thus bringing unity and completeness to the defense-in-depth concept. Taking a similar action, within reasonable and practical bounds appropriate to operating plants, would do the same for operating reactors.

The Task Force therefore concludes that the future regulatory framework should be based on the defense-in-depth philosophy, supported and modified as necessary by state-of-the-art PRA techniques. The Task Force also concludes that the application of defense-in-depth should be strengthened by formally establishing, in the regulations, an appropriate level of defense-in-depth to address requirements for "extended" design-basis events. Many of the elements of such regulations already exist in the form of the SBO rule (10 CFR 50.63), ATWS rule (10 CFR 50.62), maintenance rule (10 CFR 50.65), AIA rule (10 CFR 50.150), the requirements for protection for beyond-design-basis fires and explosions (10 CFR 50.54(hh)), and the alternative treatment requirements (10 CFR 50.69) and new reactor policy regarding regulatory treatment of nonsafety systems as described in SECY-94-084, "Policy and Technical Issues Associated with the Regulatory Treatment of Nonsafety Systems in Passive Plant Designs," dated March 28, 1994. Other elements such as SAMGs exist in voluntary industry initiatives. The Task Force envisions this collection of beyond-design-basis requirements as a coherent whole in a separate section of 10 CFR Part 50 (e.g., 10 CFR 50.200, 10 CFR 50.201) or as a dedicated appendix to 10 CFR Part 50. This separate section would have an appropriate set of quality standards, analogous to Appendix B to 10 CFR Part 50, plus a change process similar to the 10 CFR Part 52 "50.59-

like" process codified in the rule for each certified design (e.g., 10 CFR Part 52, Appendix A, Section VIII, "Processes for Changes and Departures").

The Task Force envisions a framework in which the current design-basis requirements (i.e., for anticipated operational occurrences and postulated accidents) would remain largely unchanged and the current beyond-design-basis requirements (e.g., for ATWS and SBO) would be complemented with new requirements to establish a more balanced and effective application of defense-in-depth.

This framework, by itself, would not create new requirements nor eliminate any current requirements. It would provide a more coherent structure within the regulations to facilitate Commission decisions relating to what issues should be subject to NRC requirements and what those requirements ought to be. The Task Force envisions that implementation of a new regulatory framework would result in the addition of some currently unregulated issues (i.e., those important to defense-in-depth) to a category of "extended design-basis" requirements. The framework could also support shifting issues currently addressed as design-basis requirements to the "extended design-basis" category of requirements. Such changes would establish a more logical, systematic, and coherent set of requirements addressing defense-in-depth.

As discussed earlier, the Task Force believes that voluntary industry initiatives could play a useful and valuable role in the suggested framework. Voluntary industry initiatives would not serve as substitutes for regulatory requirements but as a mechanism for facilitating and standardizing implementation of such requirements. The development of symptom-based emergency operating procedures (EOPs) in the 1980s and development of the EDMGs following the events of September 11, 2001, are just two examples of notable industry contributions to effective implementation of regulatory initiatives.

A newly organized part of the regulations would not only strengthen mitigation of accidents as severe as the Fukushima accident, but could also help to resolve some longstanding regulatory issues. For example, it could provide a natural location for requirements such as reactor coolant system breaks beyond the transition break size (being considered in the revision to 10 CFR 50.46, "Acceptance Criteria for Emergency Core Cooling Systems for Light-Water Nuclear Power Reactors"). It would also be a logical location for a requirement for PRAs and IPE and IPEEE generic safety insights for operating reactors, should the Commission desire such a requirement.

In a new regulatory framework, risk assessment and defense-in-depth would be combined more formally. PRA would help ensure that the design-basis requirements address events of a specific frequency with strict quality standards (i.e., 10 CFR Part 50, Appendix B) and that beyond-design-basis requirements address less frequent, but nonetheless important, events through appropriate quality standards. The Task Force concludes that the new framework could be implemented on the basis of full-scope Level 1 core damage assessment PRAs and Level 2 containment performance assessment PRAs.

The current NRC approach to land contamination relies on preventing the release of radioactive material through the first two levels of defense-in-depth, namely protection and mitigation. Without the release of radioactive material associated with a core damage accident, there would be no significant land contamination. The Task Force also concludes that the NRC's current approach to the issue of land contamination from reactor accidents is sound. The Task Force's objective of ensuring that protection and mitigation provide strong and effective levels of defense-in-depth is therefore fully consistent with reducing the likelihood of land contamination without introducing any new safety concepts or methods. The Task Force also considered the value of requiring a Level 3 PRA (i.e., a probabilistic assessment of accident dose and health effects) as part of a new regulatory framework. The Task Force concluded that for large light-water reactors, the metrics of core damage frequency and large early release provide very effective, relatively simple, well-documented and understood measures of safety for decisionmaking. Therefore, the Task Force has not recommended including Level 3 PRA as a part of a regulatory framework. However, some limited Level 3 PRA analyses, such as those done for a few selected plants and reported in NUREG1150, "Severe Accident Risks: An Assessment of Five U.S. Nuclear Power Plants," issued December 1990, could confirm that the selected frequency ranges for design-basis and beyond-design-basis requirements are consistent with the Commission's safety goals.

The Task Force considered the role of the IPEs and IPEEEs (mentioned above) in a new regulatory framework. If the new regulatory framework had been in place before the start of the IPE and IPEEE programs, they would likely have been different, perhaps being required rather than encouraged. However, the programs have been completed, the NRC reviewed the efforts to some extent, and the licensees have taken voluntary actions, in some cases, to address identified vulnerabilities. Given this reality, the Task Force concludes that the most appropriate step consistent with the new risk-informed, defense-in-depth regulatory framework would be to revisit the results of those efforts to identify any significant items of a generic nature and consider them for possible inclusion under design-basis or extended design-basis regulations. Similarly, this effort could identify significant plant-specific items as candidates for plant-specific regulatory requirements (e.g., license conditions).

Finally, a new and dedicated portion of the regulations would allow the Commission to recharacterize its expectations for safety features beyond design basis more clearly and more positively as "extended design-basis" requirements. The Task Force recognizes fully that a comprehensive reevaluation and restructuring of the regulatory framework would be no small feat. The Task Force also recognizes that strengthening the roles of defense-in-depth and risk assessment, emphasizing beyond-design-basis and severe accident mitigation, and establishing a clear, coherent, and well-integrated regulatory framework would be a significant accomplishment. Therefore, the Task Force concludes that additional steps would be prudent to further enhance the NRC regulatory framework to encompass the protections for accidents beyond the design basis.

Recommendation 1

The Task Force recommends establishing a logical, systematic, and coherent regulatory framework for adequate protection that appropriately balances defense-in-depth and risk considerations.

The Task Force recommends that the Commission direct the staff to initiate action to enhance the NRC regulatory framework to encompass beyond-design-basis events and their oversight through the following steps:

1.1 *Draft a Commission policy statement that articulates a risk-informed defense-in-depth framework that includes extended design-basis requirements in the NRC's regulations as essential elements for ensuring adequate protection.*

1.2 *Initiate rulemaking to implement a risk-informed, defense-in-depth framework consistent with the above recommended Commission policy statement.*

1.3 *Modify the Regulatory Analysis Guidelines to more effectively implement the defense-in-depth philosophy in balance with the current emphasis on risk-based guidelines.*

- The Task Force believes that the Regulatory Analysis Guidelines could be modified by implementing some of the concepts presented in the technology-neutral framework (NUREG-1860) to better integrate safety goals and defense-in-depth.

1.4 *Evaluate the insights from the IPE and IPEEE efforts as summarized in NUREG-1560, "Individual Plant Examination Program: Perspectives on Reactor Safety and Plant Performance," issued December 1997, and NUREG-1742, "Perspectives Gained from the Individual Plant Examination of External Events (IPEEE) Program," issued April 2002, to identify potential generic regulations or plant-specific regulatory requirements.*

Subsequent sections of this report will evaluate issues relevant to the Fukushima accident and will propose recommendations to strengthen associated NRC programs and requirements. The objective of strengthening those programs and requirements is to make them fully effective in providing defense-in-depth, thus supporting this recommended framework.

This page intentionally left blank

4. SAFETY THROUGH DEFENSE-IN-DEPTH

The key to a defense-in-depth approach is creating multiple independent and redundant layers of defense to compensate for potential failures and external hazards so that no single layer is exclusively relied on to protect the public and the environment. In its application of the defense-in-depth philosophy, the Task Force has addressed protection from design-basis natural phenomena, mitigation of the consequences of accidents, and EP.

4.1 ENSURING PROTECTION FROM EXTERNAL EVENTS

The first level of defense-in-depth is protection. Specifically, the Task Force examined the historical development of facility design bases for protection from external hazards that could cause the loss of large areas of a plant. The combined effects of the Great East Japan Earthquake of 2011 and the ensuing tsunami at the Fukushima Nuclear Power Plant site represent the most significant external event challenges that any commercial nuclear reactor has ever faced. The following section discusses the importance of establishing appropriate protection from design-basis natural phenomena. In evaluating protection from design-basis natural phenomena, the Task Force considered earthquakes, floods, high winds (due to hurricanes or tornadoes), and external fires. The Task Force concluded that earthquakes and flooding hazards warranted further Task Force consideration due, in part, to significant advancements in the state of knowledge and the state of analysis in these areas in the time period since the operating plants were sited and licensed. In addition, the earthquake and subsequent tsunami highlighted the need to evaluate concurrent related events, such as seismically induced fires and floods.

4.1.1 *Protection from Design-Basis Natural Phenomena*

BACKGROUND

The NRC has long recognized the importance of protection from natural phenomena as a means to prevent core damage and to ensure containment and spent fuel pool integrity. The NRC established several requirements addressing natural phenomena in 1971 with GDC 2, "Design Bases for Protection Against Natural Phenomena," of Appendix A to 10 CFR Part 50. GDC 2 requires, in part, that SSCs important to safety be designed to withstand the effects of natural phenomena such as floods, tsunami, and seiches without loss of capability to perform their safety functions. GDC 2 also requires that design bases for these SSCs reflect (1) appropriate consideration of the most severe of the natural phenomena that have been historically reported for the site and surrounding region, with sufficient margin for the limited accuracy and quantity of the historical data and the period of time in which the data have been accumulated, (2) appropriate combinations of the effects of normal and accident conditions with the effects of the natural phenomena, and (3) the importance of the safety functions to be performed.

Since the establishment of GDC 2, the NRC's requirements and guidance for protection from seismic events, floods, and other natural phenomena have continued to evolve. The agency has developed new regulations, new and updated regulatory guidance, and several regulatory programs aimed at enhancements for previously licensed reactors.

In 1973, Appendix A, "Seismic and Geologic Siting Criteria for Nuclear Power Plants," to 10 CFR Part 100 was established to provide detailed criteria to evaluate the suitability of proposed sites and the suitability of the plant design basis established in consideration of the seismic and geologic characteristics of the proposed sites.

In 1977, the NRC initiated the Systematic Evaluation Program (SEP) to review the designs of older operating nuclear reactor plants in order to reconfirm and document their safety. The purpose of the review was to provide (1) an assessment of the significance of differences between then-current technical positions on safety issues and those that existed when a particular plant was licensed, (2) a basis for deciding how these differences should be resolved in an integrated plant review, and (3) a documented evaluation of plant safety. The plants selected for SEP review included several that were licensed before a comprehensive set of licensing criteria (i.e., the GDC) had been developed or finalized. The SEP covered topics including seismic events, floods, high winds, and tornadoes.

In 1980, the NRC was concerned that licensees had not conducted the seismic qualification of electrical and mechanical equipment in some older nuclear reactor plants in accordance with the licensing criteria for the seismic qualification of equipment acceptable at that time. As a result, the NRC established the Unresolved Safety Issue (USI) A-46, "Seismic Qualification of Mechanical and Electrical Equipment in Operating Nuclear Power Plants," program in December 1980. In February 1987, the agency issued Generic Letter (GL) 87-02, "Verification of Seismic Adequacy of Mechanical and Electrical Equipment in Operating Reactors, Unresolved Safety Issue (USI) A-46," to address this concern. The objective of USI A-46 was to develop alternative seismic qualification methods and acceptance criteria that could be used to assess the capability of mechanical and electrical equipment in operating nuclear power plants to perform their intended safety functions. The scope of the review was limited to equipment required to bring each affected plant to hot shutdown and maintain it for a minimum of 72 hours.

In 1991, the NRC issued Supplement 4 to GL 88-20, "Individual Plant Examination of External Events (IPEEE) for Severe Accident Vulnerabilities, 10 CFR 50.54(f)." This GL requested that "each licensee perform an individual plant examination of external events to identify vulnerabilities, if any, to severe accidents and report the results together with any licensee determined improvements and corrective actions to the Commission." The external events considered in the IPEEE program include seismic events, internal fires, high winds, and floods. The primary goal of the IPEEE program was for each licensee to identify plant-specific vulnerabilities to severe accidents, if any, and to report the results, with any licensee-proposed improvements and corrective actions, to the NRC.

In 1996, the NRC established two new seismic regulations for applications submitted on or after January 10, 1997. These regulations were not applied to existing reactors. The first regulation, 10 CFR 100.23, "Geologic and Seismic Siting Criteria," sets forth the principal geologic and seismic considerations that guide the Commission in its evaluation of the suitability of a proposed site and adequacy of the design bases established in consideration of the geologic and seismic characteristics of the proposed site. The second regulation, Appendix S, "Earthquake Engineering Criteria for Nuclear Power Plants," to 10 CFR Part 50, requires that nuclear power plants be designed so that certain SSCs remain functional if the safe shutdown earthquake (SSE) ground motion occurs. These plant features are those necessary to ensure (1) the integrity of the reactor coolant pressure boundary, (2) the capability to shut down the reactor and maintain it in a safe shutdown condition, or

(3) the capability to prevent or mitigate the consequences of accidents that could result in potential offsite exposures comparable to the guideline exposures of 10 CFR 50.34(a)(1) or 10 CFR 100.11, "Determination of Exclusion Area, Low Population Zone, and Population Center Distance."

In 1996, the staff also established a new requirement in 10 CFR 100.20, "Factors To Be Considered When Evaluating Sites," for the evaluation of the nature and proximity of man-related hazards, such as dams, for applications submitted on or after January 10, 1997. This regulation was not applied to existing reactors.

In 1975, the NRC published the Standard Review Plan (SRP) (NUREG/75-087, later published as NUREG-0800, "Standard Review Plan for the Review of Safety Analysis Reports for Nuclear Power Plants: LWR Edition"), which provides standardized review criteria to assist the staff in evaluating safety analysis reports submitted by license applicants. Since its first publication, the SRP has undergone several revisions to incorporate new developments in design and analysis technology. Since the last SRP update in 2007, the staff has established interim staff guidance (ISG) in three areas related to protection from natural phenomena: (1) DC/COL-ISG-1, "Interim Staff Guidance on Seismic Issues of High Frequency Ground Motion," (2) DC/COLISG7, "Assessment of Normal and Extreme Winter Precipitation Loads on the Roofs of Seismic Category I Structures," and (3) DC/COL-ISG-20, "Seismic Margin Analysis for New Reactors Based on Probabilistic Risk Assessment." This interim guidance has been applied only to new reactor reviews.

The staff has also published several regulatory guides (RGs) that address specific technical issues related to protection from natural phenomena. These documents provide guidance to licensees and applicants on implementing specific parts of the NRC's regulations, techniques used by the NRC staff in evaluating specific problems or postulated accidents, and data needed by the staff in its review of applications for permits or licenses. These guides include the following:

- RG 1.29, "Seismic Design Classification," issued in 1972 and updated in 1973, 1976, 1978, and 2007

- RG 1.59, "Design Basis Floods for Nuclear Power Plants," issued in 1973 and updated in 1976 and 1977

- RG 1.60, "Design Response Spectra for Seismic Design of Nuclear Power Plants," issued in 1973

- RG 1.102, "Flood Protection for Nuclear Power Plants," issued in 1975 and updated in 1976

- RG 1.125, "Physical Models for Design and Operation of Hydraulic Structures and Systems for Nuclear Power Plants," issued in 1977 and updated in 1978 and 2009

- RG 1.208, "A Performance-Based Approach To Define the Site-Specific Earthquake Ground Motion," issued in 2007

The NRC staff continually evaluates new information regarding natural phenomena, including operational experience, and its potential impact on risk and overall plant safety. These evaluations have led to new requirements or guidance as discussed above, updated regulatory guidance, generic communications, and plant-specific actions to address identified issues. Several examples are presented below.

Following the Sumatra earthquake and its accompanying tsunami in December 2004, the NRC staff initiated a study to examine tsunami hazards at nuclear power plant sites, to review offshore and onshore modeling of tsunami waves, to describe the effects of tsunami waves on nuclear power plant SSCs, to develop potential approaches for screening sites for tsunami effects, to identify the repository of historic tsunami data, and to examine ways for an NRC reviewer to approach site safety assessment for a tsunami. The study, NUREG/CR-6966, "Tsunami Hazard Assessment at Nuclear Power Plant Sites in the United States of America," was published March 2009. The results of this study were incorporated in the 2007 update of SRP Section 2.4.6, "Probable Maximum Tsunami Hazards." As discussed in NUREG/CR-6966, the 1977 revision to RG 1.59 (Revision 2) was expected to include guidance for assessment of tsunamis as a flooding hazard, but that effort was not completed. The staff is in the process of updating RG 1.59 to address tsunamis and other advances in flooding analysis. Since 1977, flood estimation techniques have significantly improved with the availability of more accurate datasets and newer hydrologic, hydraulic, and hydrodynamic models. It should be noted that the current fleet of reactors was sited before RG 1.59, Revision 2, was issued.

In August 2010, the NRC initiated a proposed generic issue (GI) regarding flooding of nuclear power plant sites following upstream dam failures. The staff evaluation of this issue is ongoing.

Lastly, the NRC is evaluating seismic hazards based on new Electric Power Research Institute models used to estimate earthquake ground motion and updated models for earthquake sources in the Central and Eastern United States. The NRC is addressing this issue through the ongoing evaluation of GI-199, "Implications of Updated Probabilistic Seismic Hazard Estimates in Central and Eastern United States on Existing Plants," initiated June 9, 2005. The results of the GI-199 safety/risk assessment stage were summarized in Information Notice 2010-018, "Implications of Updated Probabilistic Seismic Hazard Estimates in Central and Eastern United States on Existing Plants," dated September 2, 2010. As discussed in Information Notice 2010-018, currently available seismic data and models show increased seismic hazard estimates for some operating nuclear power plant sites in the Central and Eastern United States. Determination of site-specific seismic hazards and associated plant risk are planned for the next phase of GI-199.

TASK FORCE EVALUATION

Current NRC regulations and associated regulatory guidance provide a robust regulatory approach for evaluation of site hazards associated with natural phenomena. However, this framework has evolved over time as new information regarding site hazards and their potential consequences has become available. As a result, the licensing bases, design, and level of protection from natural phenomena differ among the existing operating reactors in the United States, depending on when the plant was constructed and when the plant was licensed for operation. Over the years, the NRC has initiated several efforts to evaluate risks and potential safety issues resulting from these differences.

The SEP, mentioned earlier, was a one-time evaluation, and integrated plant safety assessments were published in the early 1980s for each of the plants included in the SEP. The SEP covered several technical topics, including protection from natural phenomena (i.e., floods, seismic events, tornadoes, high winds). Even that reassessment was conducted before satellite imaging, Doppler radar, and well-established theories of plate tectonics were available. It is clear that our current state of knowledge far exceeds that available to decisionmakers three decades ago.

With regard to the IPEEE program, the staff performed a limited review of the IPEEE submittals to determine whether the licensees' IPEEE processes were capable of identifying and addressing severe accident vulnerabilities caused by external events. The staff published a summary of the results of the IPEEE program in NUREG-1742, "Perspectives Gained from the Individual Plant Examination of External Events (IPEEE) Program," in April 2002. However, the NRC reviews did not attempt to validate or verify the licensees' IPEEE results or the acceptability of proposed improvements. Further, the IPEEE analyses did not document the potential safety impacts of proposed improvements, and plants were not required to report completion of proposed improvements to the NRC.

The SEP, IPEEE program, USI A-46, and other regulatory initiatives, including licensing actions to address vulnerabilities, have resulted in some plant-specific safety enhancements to address the risk of external events resulting from natural phenomena. However, the staff has not undertaken a comprehensive reestablishment of the design basis for existing plants that would reflect the current state of knowledge or current licensing criteria. As a result, significant differences may exist between plants in the way they protect against design-basis natural phenomena and the safety margin provided.

With regard to seismic hazards, as discussed above, available seismic data and models show increased seismic hazard estimates for some operating nuclear power plant sites. The state of knowledge of seismic hazards within the United States has evolved to the point that it would be appropriate for licensees to reevaluate the designs of existing nuclear power reactors to ensure that SSCs important to safety will withstand a seismic event without loss of capability to perform their intended safety function. As seismic knowledge continues to increase, new seismic hazard data and models will be produced. Thus, the need to evaluate the implications of updated seismic hazards on operating reactors will recur and need to be reevaluated at appropriate intervals.

With regard to flooding hazards, the assumptions and factors that were considered in flood protection at operating plants vary. In some cases, the design basis does not consider the probable maximum flood (PMF). In other cases, the PMF is calculated differently at units colocated at the same site, depending on the time of licensing, resulting in different design-basis flood protection. The Task Force has observed that some plants have an overreliance on operator actions and temporary flood mitigation measures such as sandbagging, temporary flood walls and barriers, and portable equipment to perform safety functions. In addition, potential dam failures have been addressed inconsistently in the establishment of the design-basis flood. In some cases, emphasis was placed on dam failures coincident with seismic events, while other mechanisms for dam failures were not fully considered. Lastly, while tsunami hazards are not expected to be the limiting flood hazard for operating plants sited on the Atlantic Ocean and the Gulf of Mexico, plants in these coastal regions do not currently include an analysis of tsunami hazards in their licensing basis. Tsunami hazards have been considered in the design basis for operating plants sited on the Pacific Ocean.

The Task Force has concluded that flooding risks are of concern due to a "cliff-edge" effect, in that the safety consequences of a flooding event may increase sharply with a small increase in the flooding level. Therefore, it would be very beneficial to safety for all licensees to confirm that SSCs important to safety are adequately protected from floods.

This reevaluation should consider all appropriate internal and external flooding sources, including the effects from local intense precipitation on the site, PMF on streams and rivers, storm surges, seiches, tsunamis, and dam failures. Similar to seismic hazards, new flooding hazard data and models will be produced from time to time. Thus, there would be a continuing benefit to having operating reactors reevaluate the implications of updated flooding hazards at appropriate intervals.

Protection from natural phenomena is critical for safe operation of nuclear power plants due to potential common-cause failures and significant contribution to core damage frequency from external events. Failure to adequately protect SSCs important to safety from appropriate design-basis natural phenomena with appropriate safety margins has the potential for common-cause failures and significant consequences as demonstrated at Fukushima.

Recommendation 2

The Task Force recommends that the NRC require licensees to reevaluate and upgrade as necessary the design-basis seismic and flooding protection of SSCs for each operating reactor.

The Task Force recommends that the Commission direct the following actions to ensure adequate protection from natural phenomena, consistent with the current state of knowledge and analytical methods. These should be undertaken to prevent fuel damage and to ensure containment and spent fuel pool integrity:

2.1   *Order licensees to reevaluate the seismic and flooding hazards at their sites against current NRC requirements and guidance, and if necessary, update the design basis and SSCs important to safety to protect against the updated hazards.*

2.2   *Initiate rulemaking to require licensees to confirm seismic hazards and flooding hazards every 10 years and address any new and significant information. If necessary, update the design basis for SSCs important to safety to protect against the updated hazards.*

2.3   *Order licensees to perform seismic and flood protection walkdowns to identify and address plant-specific vulnerabilities and verify the adequacy of monitoring and maintenance for protection features such as watertight barriers and seals in the interim period until longer term actions are completed to update the design basis for external events.*

4.1.2   *Protection from Concurrent Related Events*

BACKGROUND

The Task Force evaluated various related concurrent events and determined that fires and internal floods induced by design-basis earthquakes warranted further Task Force consideration.

Seismically induced fires are frequent after earthquakes in urban areas. Seismic events have also resulted in fires at nuclear power plants. Seismically induced fires have the potential to cause multiple failures of safety-related systems and could create fires in multiple locations at the site. Fire protection systems are not required to be functional after a seismic event; therefore, efforts to fight seismically induced fires may be impaired by degraded fire protection equipment. A seismic event may also impede offsite fire crews from reaching the site, further challenging the capability to respond to such an event.

This scenario occurred following the July 16, 2007, magnitude 6.6 Niigata-Chuetsu Oki earthquake that occurred 19 kilometers from the Kashiwazaki-Kariwa nuclear power plant, located in Niigata, Japan. Following the earthquake, a fire occurred in the Kashiwazaki-Kariwa Unit 3 electrical transformer. Sparks from a short circuit caused by large ground displacements of the transformer foundation caused the fire. The sparks ignited oil leaked from the transformer. Damage to the onsite fire protection equipment resulting from the seismic event included multiple failures of the firefighting water system in Units 1, 2, 3, and 4; failure of one of the fire water storage tanks; and failure of other fire suppression systems. Attempts by the plant fire brigade to extinguish the fire were unsuccessful. The local municipality fire brigade eventually extinguished the fire approximately 2 hours after it began. The fire was contained by fire protection walls and did not affect any plant safety equipment. However, the event provided important insights into vulnerabilities from seismically induced fires.

The 2007 Japanese earthquake event also revealed insights regarding seismically induced flooding. The plants experienced flooding from sloshing of the spent fuel pool, fire suppression piping failure outside the Unit 1 reactor building that flowed into the plant through cable penetrations, and a condenser flexible connection failure. While there were no safety consequences, these flooding failures led to water flow to various portions of the plant that could have caused SSC functional failures.

TASK FORCE EVALUATION

The staff initiated Generic Safety Issue (GSI)-172, "Multiple System Responses Program (MSRP)," to address 21 potential safety concerns that were raised by the Advisory Committee on Reactor Safeguards (ACRS) during the resolution of USI A-17, "Systems Interactions in Nuclear Power Plants"; USI A-46, "Seismic Qualification of Equipment in Operating Plants"; and USI A-47, "Safety Implications of Control Systems." GSI-172 included the ACRS concern that the resolution of USI A-46, other seismic requirements, or fire protection regulations did not adequately address seismically induced fires. This concern was identified as Item 7.4.16 in NUREG/CR-5420, "Multiple System Responses Program—Identification of Concerns Related to a Number of Specific Regulatory Issues," published October 1989. ACRS was also concerned that previous internal flooding studies had examined events such as pipe ruptures (and subsequent flooding) as single events and that the nature of a seismic event could cause such problems in multiple locations simultaneously. This concern was identified as Item 7.4.18 in NUREG/CR-5420.

The staff developed guidance for the review of the safety concerns of GSI-172 in the IPE and IPEEE programs. As a result, the IPEEE program subsumed the issues related to seismically induced fires and floods.

With regard to seismically induced fires, NUREG-1742 states the following:

> All of the IPEEE submittals reported that the licensees qualitatively examined seismically induced fire interaction issues as part of the treatment of Sandia fire risk scoping study issues. A few licensees performed a PRA study for seismically induced fire-initiating events; albeit the level of detail varied from a simplistic probabilistic analysis to inclusion in their plant's seismic or fire PRA.

> In most of the submittals, licensees included seismically induced fire considerations within the scope of their overall seismic walkdown. The level of effort, scope, and detail directed toward addressing seismically induced fire issues varied significantly

among the IPEEE submittals. One licensee did not discuss seismically induced fire evaluations in their IPEEE submittal. In most other cases, licensees limited their seismically induced fire evaluations exclusively to assessing direct impacts on safe shutdown equipment.

Seismically induced flooding events can potentially cause multiple failures of safety-related systems. The rupture of small piping could provide flood sources with the potential to affect multiple safety-related components simultaneously. Similarly, nonseismically qualified tanks are a potential flood source of concern. While some licensees proposed plant improvements to address related issues, NUREG-1742 states that the level of effort, scope, and detail directed toward addressing seismically induced flooding issues varied significantly among the IPEEE submittals. Some plants did not provide any information in their IPEEE submittals to verify this issue.

The GSI-172 issue regarding seismically induced fires and floods was closed based on the IPEEE results, and the NRC established no new requirements to prevent or mitigate seismically induced fires or floods. The Task Force concludes that the agency should reevaluate the closure of GSI-172 in light of the plant experience at the Kashiwazaki-Kariwa nuclear plant and the potential for common-mode failures of plant safety equipment as the result of seismically induced fires and floods.

Recommendation 3

The Task Force recommends, as part of the longer term review, that the NRC evaluate potential enhancements to the capability to prevent or mitigate seismically induced fires and floods.

4.2   MITIGATION

The second level of defense-in-depth is mitigation. The Great East Japan Earthquake of 2011 and the ensuing tsunami resulted in many mitigation systems at the Fukushima Dai-ichi Nuclear Power Plant being unable to operate. The subsequent challenges faced by the operators at Fukushima Dai-ichi were beyond any faced previously at a commercial nuclear reactor. The Task Force examined the U.S. regulations, guidance, and practices for mitigating the consequences of accidents similar to those that occurred at Fukushima Dai-ichi. The following sections discuss the Task Force evaluation of insights from Fukushima and provide recommendations for enhancing the mitigation capability of U.S. reactors with regard to prolonged loss of ac power, containment overpressure protection, combustible gas control, spent fuel pool safety, and onsite emergency actions.

4.2.1   *Prolonged Loss of Alternating Current Power*

BACKGROUND

Alternating current (ac) electrical power is critically important to the safety of nuclear power plants. Many of the SSCs intended to cool the nuclear fuel in the reactor and in the spent fuel pools, to maintain radioactive containment systems, and to provide ventilation systems to minimize release of radioactive materials rely on ac power. These systems depend on electrical power to drive pumps, fans, and compressors, operate instrumentation and control systems, and run motors to open and close valves and dampers. For these reasons, the loss of all ac power both onsite and offsite, as occurred at Fukushima, is highly significant. With this in mind, the Task Force critically evaluated the design-basis protections to prevent loss of ac power, as discussed above, and the plants' ability to maintain safety functions following the loss of ac power as discussed below.

The NRC SBO rule (10 CFR 50.63) requires that each nuclear power plant must be able to cool the reactor core and maintain containment integrity for a specified duration of an SBO (defined in 10 CFR 50.2, "Definitions," as a complete loss of required onsite and offsite ac electrical power). The specified duration is based on the following factors:

- the redundancy of the onsite emergency ac power sources
- the reliability of the onsite emergency ac power sources
- the expected frequency of loss of offsite power
- the probable time needed to restore offsite power

RG 1.155, "Station Blackout," describes an acceptable means to comply with 10 CFR 50.63. It primarily addresses three areas: (1) maintaining highly reliable onsite ac electric power systems, (2) developing procedures and training to restore offsite and onsite emergency ac power should either one or both become unavailable, and (3) ensuring that plants can cope with an SBO for some period of time based on the probability of occurrence of an SBO, as well as the capability for restoring ac power to the site in a timely fashion. The RG provides an acceptable method for determining the specified duration for withstanding an SBO considering the four factors identified in the rule language (10 CFR 50.63(a)(1)(i)–(iv)). The method described in RG 1.155 results in a minimum acceptable SBO duration capability ranging from 2 to 16 hours. The result for all operating plants was a coping duration of 4 to 8 hours.

In evaluating the expected frequency of loss of offsite power, the guide addresses the expected frequency of high winds, including those from tornadoes and hurricanes, and the expected annual snowfall. The impact of other external hazards (e.g., seismic and flooding) on the frequency of loss of offsite power is not addressed. Nor does the guide address the concurrent consequences on the facility of the external hazards impacting offsite power. Consequently, common-cause failures of all onsite and offsite power resulting from a naturally occurring external event are not considered.

The analysis supporting the requirements of 10 CFR 50.54(hh) demonstrates, among other things, the potential for prolonged SBO conditions arising from terrorist actions including large aircraft impacts and high explosives.

In a letter from J.E. Dyer (NRC) to A. Pietrangelo (Nuclear Energy Institute (NEI)) dated December 22, 2006, the NRC endorsed industry-developed guidance (NEI 06-02, Revision 0, "License Amendment Request (LAR) Guidelines," issued December 2006) to demonstrate compliance with the requirements reflected in the NRC's 2002 Interim Compensatory Measures Order (and, subsequently, in 10 CFR 50.54(hh)(2)). To comply with the requirements, each nuclear power plant licensee expanded command and control capabilities and developed strategies and capability to provide cooling to fuel in the reactor and spent fuel pool and to mitigate releases without reliance on the site's ac electrical power distribution system. The following mitigation capabilities were addressed:

- for pressurized water reactors (PWRs):
  - » additional sources of coolant water for the reactor and steam generators

> » methods to reduce pressure in and feed cooling water to the steam generators

> » methods to cool the reactor core and minimize releases of radioactive materials from containment

- for BWRs:

> » additional sources of coolant water for the reactor

> » methods to reduce reactor pressure and feed cooling water to the reactor

> » methods to cool the reactor core and reduce releases of radioactive materials from containment

- for spent fuel pools:

> » additional sources of coolant water for and methods to inject or spray to the spent fuel pools

> » methods to control leakage from damage to the spent fuel pools

> » methods to reduce releases of radioactive materials from the spent fuel pools

The strategies, called EDMGs, implemented to meet the requirements of the 2002 Interim Compensatory Measures Order, subsequent facility-specific license conditions, and ultimately 10 CFR 50.54(hh)(2), did not address external natural hazards (e.g., seismic, flooding, tornadoes, hurricanes) or initiating events other than extensive damage to the facilities caused by large fires or explosions. The EDMGs were designed to assist in the mitigation of one particular beyond-design-basis scenario (i.e., loss of large areas of the facility due to fires or explosions) that typically could involve SBO. The implementing guidance for the EDMGs sets expectations that the amount and capacity of equipment to implement the strategies should be sufficient to mitigate the consequences of an event that affects only one unit. In addition, the implementing guidance for the EDMGs specifies that the required mitigation equipment used under the EDMGs should be stored in an area physically separated by more than 91 meters (300 feet) from the equipment that could have been damaged by a large fire or explosion. The equipment is not expected to be protected from design-basis or beyond-design-basis external events, such as floods, earthquakes, or high winds.

Finally, critical instrumentation and control systems typically depend on the availability of direct current (dc) electrical power at the facility. During a prolonged SBO, when ac power would not be available and the battery banks become depleted, functional failure would occur for nearly all instrumentation and control systems for monitoring critical parameters and operating critical systems that ensure the integrity of the fuel in the reactor and spent fuel pools and maintaining containment structures.

TASK FORCE EVALUATION

Information available at the time of this report indicates that the earthquake at Fukushima Dai-ichi on March 11, 2011, caused a loss of all offsite sources of ac power to the six units, and the ensuing tsunami caused failure of the emergency diesel generators for Units 1 through 4. It appears that Unit 1 also suffered damage to its onsite dc power source because of the tsunami. While most of the emergency diesel generators in Units 5 and 6 also failed due to the tsunami, Unit 6 had one air-cooled emergency diesel generator that was not lost; operators cross-tied that generator to essential Unit 5 and 6 systems, thus retaining the ability to cool the Unit 5 and 6 reactors and spent fuel pools. The scope of the damage to the offsite power infrastructure from the earthquake combined with the damage to the site from the tsunami resulted in the inability to restore any ac electrical power to Units 1 through 4 for many days. The onsite emergency ac power sources have not yet been restored and, while ac power from offsite has been rerouted to the facility, damage to the onsite electrical distribution and other critical equipment from the tsunami has not allowed full utilization of this offsite ac power for installed plant equipment in Units 1 through 4. These four units were in a prolonged SBO for many days and still are significantly challenged regarding the reliable distribution and use of ac electrical power.

Under these circumstances, factors that prevented restoration of power included (1) common-cause failure of onsite emergency ac sources due to flooding, (2) common-cause failure of ac electrical distribution due to flooding, (3) common-cause failure of offsite (nonemergency) ac power distribution due to earthquake ground motion, and (4) offsite infrastructure degradation from the earthquake and tsunami, which may have impeded efforts to restore offsite power.

The Commission's SBO requirements provide assurance that each nuclear power plant can maintain adequate core cooling and maintain containment integrity for its approved coping period (typically 4 or 8 hours) following an SBO. Also, if available, the equipment used for compliance with 10 CFR 50.54(hh)(2) would provide additional ability to cool either the core or the spent fuel pool and mitigate releases from primary and secondary containment during a prolonged SBO. The implementing guidance for SBO focuses on high winds and heavy snowfalls in assessing potential external causes of loss of offsite power, but does not consider the likelihood of loss of offsite power from other causes such as earthquakes and flooding. Also, the SBO rule does not require the ability to maintain reactor coolant system integrity (i.e., PWR reactor coolant pump seal integrity) or to cool spent fuel. Further, the SBO rule focuses on preventing fuel damage and therefore does not consider the potential for the buildup of hydrogen gas inside containment during a prolonged SBO condition and the potential need to power hydrogen igniters in certain containment designs to mitigate the buildup of hydrogen. Nor does it consider containment overpressure considerations and the need to vent the containment in certain designs. Finally, the SBO rule does not require consideration of the impact on the station, and particularly on the onsite ac generation and distribution, of the natural event that caused the loss of offsite ac electrical power.

During the prolonged SBO condition at the Fukushima Dai-ichi units, after the batteries were damaged or depleted and no ac power was available to operate equipment or recharge the batteries, the operators faced significant challenges in understanding the condition of the reactors, containments, and spent fuel pools because instrumentation was either lacking or not functioning properly.

The Task Force concludes that revising 10 CFR 50.63 to expand the coping capability to include cooling the spent fuel, preventing a loss-of-coolant accident, and preventing containment failure would be a significant benefit. The Task Force also concludes that a strategy is needed to provide these functions for a prolonged period without ac power from the normal offsite or emergency onsite sources without the vital ac distribution systems within the plant. The Task Force developed a three-part strategy to achieve this objective. First, licensees would need a coping capability to maintain these functions for at least 8 hours at each unit. This capability should minimize reliance on operator action during this period in recognition of the potential for adverse work conditions related to the cause of the prolonged SBO. Also, this 8-hour coping period would provide sufficient time to enable the operators to focus their efforts on restoring ac power and deploying onsite equipment for the extended coping period to follow. Second, licensees would need an extended coping capability to maintain these functions for at least 72 hours. The Task Force envisions that establishing this extended coping capability would involve extensive operator actions during the 8-hour coping period to deploy portable equipment maintained at the site in a manner that protects it from the severe natural phenomena that may cause it to be needed. The purpose of the extended coping capability is to provide a bridge to the third component, which would be a sustainable cooling capability that would utilize preplanned and prestaged equipment at an offsite location that could be delivered and installed within 72 hours. The planning would need to consider the potential for degraded transportation infrastructure resulting from severe natural phenomena. The Task Force envisions that several sites might share the same prestaged equipment, provided that those sites are not susceptible to severe damage from the same initiating event (e.g., earthquake or hurricane) requiring concurrent demand for the same equipment.

To achieve the goal of providing an effective level of defense-in-depth for SBOs caused by external events beyond the design basis, the SBO mitigation equipment would need to be protected from such events. Such protection was not available at Fukushima, where beyond-design-basis flooding caused a prolonged SBO.

In the Task Force's proposed risk-informed, defense-in-depth framework (Recommendation 1 in Section 3 of this report), the extent of the beyond-design-basis external event that would require effective SBO mitigation would be established based on the likelihood (i.e., estimated frequency) of such events. Extensive work on seismic events has demonstrated that significant margin exists beyond a well-formulated seismic design-basis SSE. Typically, a margin of 2 exists above an SSE. That is to say, plant equipment failures are unlikely unless seismic loads are about twice the design-basis SSE loads. Such loads generally correspond to earthquake frequencies 5 to 10 times less likely than the design-basis SSE. The Task Force therefore concludes that SBO equipment designed to the SSE would likely be sufficiently robust to function following a reasonably foreseeable beyond-design-basis seismic event.

The Task Force has concluded that the situation is somewhat different in terms of beyond-design-basis flooding. First, flooding can be caused by a number of different phenomena: river flooding; dam failure; precipitation; storm surge; tsunami; or internal failures of pipes, pumps, or tanks within the plant. Second, flooding can have a cliff-edge effect; that is, a small increase in flooding level can produce a large effect in terms of equipment failure and potential plant damage. With respect to this issue, the experience at Fukushima Dai-ichi Units 5 and 6 appears more informative than that at Fukushima Dai-ichi Units 1, 2, 3, and 4. Units 5 and 6 are sited at an elevation of 13 meters (43 feet) above sea level and, based on the information available, the tsunami reached a level of 14 meters (46 feet), producing about 1 meter (3 feet) of flooding on the site. In contrast, Units 1 through 4 appear to have been inundated by about 5 meters (16 feet) of sea water. The extensive damage at Units 1 through 4 is therefore not surprising. However, Units 5 and 6 also experienced extensive damage of critical safety equipment as a result of about 1 meter (3 feet) of flooding, leaving the units at significant risk of core damage. Only one air-cooled diesel generator remained available at Unit 6 and functioned with significant operator action to maintain cooling at the two units. This circumstance illustrates the concept of a cliff-edge effect.

The Task Force also considered external fires, as well as tornado and hurricane winds, as they relate to the qualification of SBO equipment. The Task Force has concluded that the existing regulations for both fires and high-wind requirements include sufficient safety margins to obviate the need for additional special protection for SBO equipment. The Task Force concludes that neither external fires nor high winds should be characterized as "cliff-edge" phenomena.

Based on the preceding considerations, the Task Force concludes that, to have SBO equipment function effectively as a layer of defense-in-depth, it would need to be protected from flooding beyond the design basis. The Task Force has also concluded that the safety margin built into the design-basis flood would not be sufficient to provide the desired level of protection.

The Task Force considers the issue of flood protection for SBO equipment to be significant and recognizes that flooding protection of such equipment would require it to be located at a suitable elevation (above the design-basis flood plus a significant margin) or require it to have an effective watertight enclosure. Establishing such protections may be difficult at some sites. Nevertheless, the Task Force concludes that such protection of SBO equipment is essential to implementation of the recommended framework for reactor safety.

A beyond-design-basis flood could be established through extensive, probabilistic hazards analysis. As a practical matter, and to prevent undue delays in implementing additional SBO protections, the Task Force concludes that locating SBO mitigation equipment in the plant one level above flood level (about 5 to 6 meters (15 to 20 feet)) or in watertight enclosures would provide sufficient enhanced protection for this level of defense-in-depth.

These recommendations for revision to 10 CFR 50.63 would provide additional safety margins for a prolonged SBO as a part of the overall risk-informed, defense-in-depth regulatory framework providing adequate protection of public health and safety.

In addition, the EDMGs and associated equipment could be helpful and available promptly to the operators to mitigate accidents such as those that occurred at Fukushima. However, the two issues discussed above result in limited effectiveness of the EDMG strategies for naturally occurring events that typically affect more than one unit.

As an interim measure, the equipment to implement the EDMGs would need to be reasonably protected from external events (i.e., stored in existing locations that are reasonably protected from significant floods and involve robust structures with enhanced protection from seismic and wind-related events). In addition, this equipment would need to be expanded to provide sufficient capacity to allow for a multiunit event response.

Recommendation 4

The Task Force recommends that the NRC strengthen SBO mitigation capability at all operating and new reactors for design-basis and beyond-design-basis external events.

The Task Force recommends that the Commission direct the staff to begin the actions given below to further enhance the ability of nuclear power plants to deal with the effects of prolonged SBO conditions at single and multiunit sites without damage to the nuclear fuel in the reactor or spent fuel pool and without the loss of reactor coolant system or primary containment integrity.

4.1   Initiate rulemaking to revise 10 CFR 50.63 to require each operating and new reactor licensee to (1) establish a minimum coping time of 8 hours for a loss of all ac power, (2) establish the equipment, procedures, and training necessary to implement an "extended loss of all ac" coping time of 72 hours for core and spent fuel pool cooling and for reactor coolant system and primary containment integrity as needed, and (3) preplan and prestage offsite resources to support uninterrupted core and spent fuel pool cooling, and reactor coolant system and containment integrity as needed, including the ability to deliver the equipment to the site in the time period allowed for extended coping, under conditions involving significant degradation of offsite transportation infrastructure associated with significant natural disasters.

- The purpose of the 8-hour minimum coping capability is to enable operators to restore ac power or to establish the proposed 72-hour extended coping capability. Core and spent fuel pool cooling would need to be provided and primary containment isolation capability and reactor coolant system integrity maintained by equipment (including essential instrumentation and controls) independent of ac power (e.g., turbine-driven, diesel-driven, dc-operated, air-operated, or passive systems).

- The 8-hour coping systems and equipment would be protected from damage from all design-basis events and extended beyond-design-basis events by either locating the equipment one level (i.e., 5 to 6 meters (15 to 20 feet)) above the plant design-basis flooding level or in water-tight enclosures.

- The 8-hour coping strategy should ensure that core and spent fuel pool cooling is maintained and unmanageable leakage of coolant does not occur (e.g., from PWR reactor coolant pump seal failure).

- The 8-hour coping strategy should also ensure that containment integrity can be established if needed, including the capability to operate wetwell vents for BWR facilities with Mark I and Mark II containments, and one train of hydrogen igniters at BWR facilities with Mark III containments and at PWR facilities with ice condenser containments.

- The design of the systems supporting the 8-hour minimum coping time should be based on a conservative analysis of the capacity of ac-independent equipment and instrumentation. The 8-hour coping capability should only rely on permanently installed equipment. Operator actions relied upon to maintain this coping capability during this 8-hour period should be limited to those types of actions consistent with routine types of operational activities governed by established procedures and training. This would enable the operators to focus on actions needed to restore ac power and to deploy and operate equipment to be used for the extended coping period. The systems and procedures necessary for this function need not be single failure-proof, but they should be included in the plant's licensing basis (i.e., described in the final safety analysis report) and should be subject to the quality assurance requirements of Appendix B to 10 CFR Part 50.

- The design of the systems supporting the 72-hour extended coping time should cover the same scope of functions as the 8-hour minimum coping time, but it can be based on realistic analysis with reasonable operator action using portable or permanently installed equipment governed by established procedures and training. This extended coping time will be sufficient to allow time for the effective acquisition, transportation, installation, and use of preplanned and prestaged offsite resources for continued achievement of the goals of core and spent fuel pool cooling, and reactor coolant system and primary containment integrity.

- The prestaged equipment could be used to satisfy these requirements at multiple sites as long as the sites would not reasonably be expected to experience the same natural event at the same time.

- As part of the revision to 10 CFR 50.63, the NRC should require that the equipment and personnel necessary to implement the minimum and extended coping strategies shall include sufficient capacity to provide core and spent fuel pool cooling, and reactor cooling system and primary containment integrity for all units at a multiunit facility. The staff should also make the appropriate revisions to the definitions of "station blackout" and "alternate ac source" in 10 CFR 50.2.

4.2 Order licensees to provide reasonable protection for equipment currently provided pursuant to 10 CFR 50.54(hh)(2) from the effects of design-basis external events and to add equipment as needed to address multiunit events while other requirements are being revised and implemented.

- This existing equipment currently provides some of the coping capability that is recommended for the long term, but current storage requirements do not ensure that it will be available after a design-basis external event. This requirement would increase the likelihood that the equipment will be available if called upon.

- The staff should also consider conforming changes to the requirements in 10 CFR 50.54(hh)(2) to address multiunit response capacity.

4.2.2 Containment Overpressure Protection

BACKGROUND

BWRs have robust capability to provide cooling water to the reactor core through many diverse safety-related and nonsafety-related systems when ac power is available. In addition, BWRs typically have two ac-independent ways of providing cooling water. In October 1975, the NRC published NUREG-75/014, "Reactor Safety Study: An Assessment of Accident Risks in U.S. Commercial Nuclear Power Plants (WASH-1400)." One of the many insights developed through the study was that the risk of containment failure during severe accidents was higher at BWRs with Mark I containments because the containment volume of Mark I containment designs was significantly less than that of the other containment designs, approximately one-sixth the volume of large dry PWR containments.

In December 1990, the NRC published NUREG-1150, which documented a relatively high containment failure probability in the event of core damage at a BWR with a Mark I containment design because of the smaller containment volume. The failure probability was dominated by accident sequences involving a loss of feedwater with loss of decay heat removal (due to multiple failures of safety systems) resulting in core damage and subsequent containment failure. These sequences are often referred to as "TW" sequences. A prolonged SBO can also produce these failure sequences.

Concurrent with the development of NUREG-1150, the NRC completed the Containment Performance Improvement Program for Mark I Containments, identifying a number of recommendations to enhance the performance of these containment designs. The NRC staff described the improvement program recommendations in SECY-89-017, "Mark I Containment Performance Improvement Program," issued January 1989. The Commission evaluated the staff recommendations and determined that the containment improvements should be evaluated on a plant-specific basis through the IPE program, with the exception of the recommendation for hardened wetwell vents. Specifically, the Commission directed the staff to approve the installation of hardened vents under the provisions of 10 CFR 50.59, "Changes, Tests and Experiments," for licensees that, on their own initiative, elected to incorporate that improvement into their plants.

The staff issued GL 89-16, "Installation of a Hardened Wetwell Vent," on September 1, 1989, describing the safety benefits of the hardened vent modification and requesting licensees to provide information on their plans to install that modification, or to provide cost information should they choose not to install the modification. Eventually, all BWR facilities with Mark I containment designs voluntarily installed a hardened vent. No regulatory requirement was imposed. The designs of those vents varied from plant to plant. At some facilities, the hardened vent relied on ac-operated valves, some relied on dc-operated or air-operated valves, and some incorporated passive rupture disks along with isolation valves. Each different design has different operational complexities during a prolonged SBO scenario. The updated final safety analysis report for each facility includes a description of the hardened vent, but the vent is not a required design feature for that facility.

In BWRs with a Mark II containment design, the containment volume could be approximately 25 percent larger than the volume of Mark I containments. In the resolution of GSI157, "Containment Performance," the staff concluded that that the need for hardened vents at BWRs with Mark II containments should be evaluated on a plant-specific basis through the IPE program. Eight BWR units in the United States have Mark II containment designs. Three of these units have installed hardened vents, and the remaining five units at three sites have not installed hardened vents.

TASK FORCE EVALUATION

Information available at the time of this report indicates that, during the days following the Fukushima Dai-ichi prolonged SBO event, primary containment (drywell) pressure in Units 1, 2, and 3 substantially exceeded the design pressure for the containments. The operators attempted to vent containment, but they were significantly challenged operating the wetwell (suppression pool) vents because of complications from the prolonged SBO. Units 1, 2, 3, and 4 use the Mark I containment design; however, because Mark II containment designs are only slightly larger in volume than Mark I containment designs, it can reasonably be concluded that a Mark II under similar circumstances would have suffered similar consequences.

The process at Fukushima Dai-ichi Units 1, 2, 3, and 4 for venting the wetwell involves opening one ac-powered motor-operated valve to permit air pressure to open air-operated valves in the vent line, and then opening another ac-powered motor-operated valve in line with the air-operated valves, permitting containment pressure to impact a rupture disk designed to open if containment pressure is significantly above design pressure. If all of these actions are successful, the containment would vent directly to the plant stack, and containment integrity could be reestablished by closing either the in-line ac-powered motor-operated valve or the air-operated valves. In a prolonged SBO situation, these actions would not be possible from the control room because of the loss of ac power and the depletion of the batteries providing dc control power for the valves. It is unclear whether the operators were ever successful in venting the containment in Unit 1, 2, or 3. The operators' inability to vent the containments complicated their ability to cool the reactor core, challenged the containment function, and likely resulted in the leakage of hydrogen gas into the reactor building, precipitating significant explosions in Units 1, 3, and 4.

Ensuring that BWR Mark I and Mark II containments have reliable hardened venting capability would significantly enhance the capability of those BWRs to mitigate serious beyond-design-basis accidents. A reliable venting system could be designed to be independent of ac power and to operate with limited operator actions from the control room. Alternatively, a reliable venting capability could be provided through a passive containment venting design, such as rupture disks with ac-independent isolation valves to reestablish containment following rupture of the disk. The Task Force concludes that the addition or confirmation of a reliable hardened wetwell vent in BWR facilities with Mark I and Mark II containment designs would have a significantly safety benefit.

During the longer term review, the staff needs to reevaluate the design of other containment structures for operating reactors to reaffirm the past conclusion that hardened vents are not necessary to mitigate certain beyond-design-basis accident scenarios.

Recommendation 5

The Task Force recommends requiring reliable hardened vent designs in BWR facilities with Mark I and Mark II containments.

The Task Force recommends that the Commission direct the staff to take the following actions to ensure the effectiveness of hardened vents:

5.1 *Order licensees to include a reliable hardened vent in BWR Mark I and Mark II containments.*

- This order should include performance objectives for the design of hardened vents to ensure reliable operation and ease of use (both opening and closing) during a prolonged SBO.

5.2 *Reevaluate the need for hardened vents for other containment designs, considering the insights from the Fukushima accident. Depending on the outcome of the reevaluation, appropriate regulatory action should be taken for any containment designs requiring hardened vents.*

4.2.3 Combustible Gas Control

BACKGROUND

The NRC regulations in 10 CFR 50.44, "Combustible Gas Control for Nuclear Power Reactors," require reactors either to operate with their containment atmosphere inerted, resulting in the lack of oxygen to support combustion, or to have the capability for controlling combustible gas generated from a metal-water reaction involving approximately three-quarters of the fuel cladding so that there is no impact on the containment structural integrity. RG 1.7, "Control of Combustible Gas Concentrations In Containment," provides further guidance on regulatory expectations for combustible gas control.

BWR facilities with Mark I and Mark II containment structures are required to operate their containments with inerted atmospheres. BWR facilities with Mark III containments and

PWR facilities with ice condenser containments are required to have hydrogen igniters inside containment to control the buildup of hydrogen gas. These igniters are operated in two redundant trains, with each train powered by one of the redundant safety-grade dc electrical power systems. GSI-189 raised questions about the effectiveness of these igniter systems during a prolonged SBO scenario. In response to the issues raised in GSI-189, licensees operating BWRs with Mark III containments and PWRs with ice condenser containments voluntarily installed nonsafety-grade backup electrical power to one train of the igniters, independent of the safety-grade ac and dc onsite power systems. PWR facilities with large dry containments do not control hydrogen buildup inside the containment structure because the containment volume is sufficient to keep the pressure spike of potential hydrogen deflagrations within the design pressure of the structure.

TASK FORCE EVALUATION

Information available at the time of this report indicates that, during the days following the Fukushima prolonged SBO event, Units 1, 3, and 4 experienced explosions, causing significant damage to the reactor buildings for those units. It is believed that the explosions in Units 1 and 3 resulted from hydrogen gas that was liberated inside the drywell during high-temperature zirconium fuel cladding reactions with water and that hydrogen gas migrated to the reactor building. The migration route of the hydrogen gas from the primary containment to reactor building has not yet been determined definitively; however, the failure to prevent, through containment venting, the primary containment pressure from significantly exceeding the design pressure likely contributed to the transport of hydrogen gas. It is believed that the explosion in the Unit 4 reactor building also resulted from hydrogen gas, but the source of the gas in Unit 4 is not yet clear. Unit 2 may also have experienced a hydrogen explosion in its suppression pool inside containment. However, the mechanism for suppression pool failure remains unclear.

The method of combustible gas control in BWR Mark I and Mark II containments (i.e., containment inerting with nitrogen) will prevent hydrogen fires or explosions as long as containment remains isolated, but it will not eliminate the hydrogen resulting from an accident damaging the core. In contrast, other designs eliminate hydrogen through controlled burning (BWR Mark III and PWR ice condenser containment designs) or by accommodating an associated hydrogen explosion (PWR large dry containments). This means that in a BWR Mark I or Mark II containment, the hydrogen must be kept in containment by controlling containment pressure without venting (i.e., through heat removal from the containment when possible) or by venting to a safe location.

Implementation of Task Force Recommendation 4, associated with prolonged SBO, would reduce the likelihood of core damage and hydrogen production. In addition, implementation of Recommendation 5 to enhance the containment venting capabilities for Mark I and Mark II containments, while primarily intended for overpressure protection, would also provide for the reliable venting of hydrogen to the atmosphere. These two steps would greatly reduce the likelihood of hydrogen explosions from a severe accident.

Sufficient information is not yet available for the Task Force to reasonably formulate any further specific recommendations related to combustible gas control based on insights from the Fukushima Dai-ichi accident.

Recommendation 6

The Task Force recommends, as part of the longer term review, that the NRC identify insights about hydrogen control and mitigation inside containment or in other buildings as additional information is revealed through further study of the Fukushima Dai-ichi accident.

4.2.4 Spent Fuel Pool Safety

BACKGROUND

SSCs for spent fuel storage and handling have safety classifications that reflect their importance to safety. SSCs essential to retaining the inventory of spent fuel pool water covering the spent fuel and to maintaining a substantial margin to criticality are typically classified as safety related. Such safety-related SSCs include the spent fuel pool structure and penetrations, the spent fuel storage racks, the neutron-absorbing panels in the racks, and the spent fuel itself. Some fuel handling equipment is also safety related. Because the consequences of many fuel handling events and loss of spent fuel forced cooling events have been evaluated and found to be small, these events are not classified as design-basis events. Consequently, other spent fuel storage and handling equipment and spent fuel pool water inventory makeup and cooling systems may not be classified as safety related. At U.S. reactors, some of the spent fuel pool cooling and makeup systems are powered by safety-grade ac electrical power and some are powered by nonsafety-grade ac electrical power.

Licensees developed the EDMGs, described in Section 4.2.1 of this report, to maintain or restore spent fuel pool cooling and mitigate releases under the circumstances associated with the loss of large areas of the plant due to fire or explosion. These requirements arose from analyses conducted by the NRC staff after September 11, 2001, that demonstrated, among other things, the potential for prolonged SBO conditions.

The NRC regulations in 10 CFR 50.36, "Technical Specifications," describe the requirements for technical specifications included in each facility operating license. The technical specifications describe requirements, called limiting conditions for operation (LCOs), for the characteristics of safety systems and components that must be available during various reactor operating modes. Specifically for electrical power systems, when the reactor is operating, the LCOs typically require two independent onsite and at least two offsite ac electrical power systems to be operable to provide electrical power to the safety equipment. When the reactor is shut down and defueled for maintenance work and all of the fuel is placed in the spent fuel pool, the LCOs do not require any electrical power systems to be operable. In this condition, the heat load in the spent fuel pool is highest, and the time margin to boil could be as little as several hours immediately after transfer of the fuel from the reactor to the spent fuel pool. The fuel could be at risk of uncovery in as a few as 10 hours without pool cooling; ac electrical power is important to the cooling the nuclear fuel in the spent fuel pools and to providing ventilation systems to minimize the release of radioactive materials. The regulation in 10 CFR 50.36(b) authorizes the NRC to include additional technical specifications as it finds appropriate.

The spent fuel pools at Fukushima Dai-ichi Units 1 through 4 contained many fewer assemblies than typically stored in U.S. reactor unit spent fuel pools. Unit 4 contained more assemblies than the other units because Unit 4 was in an extended outage and the reactor was defueled, transferring all assemblies to the spent fuel pool to facilitate maintenance activities associated with the reactor systems.

Fuel Assemblies in the Reactor Cores and Spent Fuel Pools at Fukushima Dai-ichi Units 1 through 4		
	Reactor Core	Spent Fuel Pool
Unit 1	400	292
Unit 2	548	587
Unit 3	548	514
Unit 4	0	1,331

The storage capacity of U.S. reactor unit spent fuel pools ranges from less than 2,000 assemblies to nearly 5,000 assemblies, with an average storage capacity of approximately 3,000 spent fuel assemblies. Typically, the U.S. spent fuel pools are filled with spent fuel assemblies up to approximately three-quarters of their capacity. In addition to the unit-specific spent fuel pools at the Fukushima Dai-ichi facility, a separate spent fuel storage building existed onsite with wet pool storage of spent fuel containing 6,291 spent fuel assemblies. U.S. reactor facilities do not typically have an additional spent fuel wet storage building like that at Fukushima Dai-ichi.

A spent fuel assembly just removed from a BWR after an operating cycle generates approximately 10 kilowatts of heat, and that heat level diminishes rapidly over time. Within a short period of time after removal from the reactor, spent fuel assemblies are coolable with water sprays, and within a few years, the assembly rate of heat generation is reduced to a few percent of its original heat generation rate. Consequently, the number of spent fuel assemblies in the spent fuel pool does not significantly affect the ability to cool the spent fuel in the pool.

Many U.S. reactors have dry spent fuel storage capability in addition to wet storage in the spent fuel pools. These dry storage facilities are called independent spent fuel storage installations, and fuel stored in such an installation first must be placed into a dry cask. The NRC reviews and certifies the designs for these dry casks. Typically, 5 years must pass from the time an assembly is removed from the reactor before it can be placed into a dry cask. After 5 years, the heat generation rate is very low.

In addition to responding to prolonged SBO conditions, the EDMGs also address contingencies for cases when the spent fuel pools are unable to retain water above the top of the fuel. This situation could also occur with significant beyond-design-basis seismic events. One effective strategy for mitigation is to blanket the spent fuel with a water spray.

Current spent fuel pool instrumentation provides limited indication and typically depends on the availability of dc electrical power at the facility. That power is provided either through inverters powered by ac electrical power or by the station's safety-grade redundant battery banks. Direct spent fuel pool level indication is rarely provided in the control room for the current nuclear fleet. Typically, level is measured using a level switch in the skimmer surge tank. During a prolonged SBO, ac power would not be available and the battery banks would be depleted, resulting in functional failure of nearly all instrumentation and control systems for monitoring spent fuel pool parameters and operating systems ensuring the integrity of the fuel in the spent fuel pools.

TASK FORCE EVALUATION

Information available at the time of this report indicates that the earthquake at Fukushima caused a loss of all offsite sources of ac power to the six units, and the ensuing tsunami caused the failure of the emergency diesel generators for Units 1 through 4. The scope of the damage to the offsite power infrastructure from the earthquake, combined with the damage to the site from the tsunami, resulted in the inability to restore any ac electrical power to Units 1 through 4 for many days.

During this protracted SBO condition, no ac power was available to operate equipment, and the batteries were depleted. This resulted in having no onsite capability to provide water inventory or cooling to the spent fuel pools, and the operators were significantly challenged in understanding the condition of the spent fuel pools because of the lack of instrumentation or because of instrumentation that was not functioning properly. Eventually, spent fuel cooling was provided by pumper trucks employing high booms to spray water from a distance into the spent fuel pools.

The reliability and availability of U.S. spent fuel pool makeup systems would be better ensured if the NRC had a requirement for those systems to have safety-related ac power that is controlled under a technical specification LCO.

Substantial additional defense-in-depth would be provided, and cooling the spent fuel in a prolonged SBO would have been substantially simplified, with an installed seismically qualified means to spray water into the spent fuel pools, including an easily accessible connection to supply the water (e.g., using a portable pump or pumper truck) at grade outside the building.

The lack of information on the conditions of the fuel in the Fukushima spent fuel pools was a significant problem in monitoring the course of the accident and contributed to a poor understanding of possible radiation releases and to confusion about the need and priorities for support equipment. The Task Force therefore concludes that reliable information on the conditions in the spent fuel pool is essential to any effective response to a prolonged SBO or other similarly challenging accident.

The Task Force concludes that clear and coherent requirements to ensure that the plant staff can understand the condition of the spent fuel pool and its water inventory and coolability and to provide reliable, diverse, and simple means to cool the spent fuel pool under various circumstances are essential to maintaining defense-in-depth. The Task Force sees significant value in ensuring that sufficient cooling capacity exists under various design-basis natural phenomena.

Final understanding of the detailed sequence of events and the condition of the spent fuel pools will not be fully developed for some time. Based on the information to date, it is clear that the two most cogent insights from the Fukushima accident related to spent fuel pool safety concern (1) the instrumentation to provide information about the condition of the pool and the spent fuel and (2) the plant's capability for cooling and water inventory management. The Task Force's recommendations address these two critical areas.

Recommendation 7

The Task Force recommends enhancing spent fuel pool makeup capability and instrumentation for the spent fuel pool.

The Task Force recommends that the Commission direct the staff to do the following:

7.1   *Order licensees to provide sufficient safety-related instrumentation, able to withstand design-basis natural phenomena, to monitor key spent fuel pool parameters (i.e., water level, temperature, and area radiation levels) from the control room.*

7.2   *Order licensees to provide safety-related ac electrical power for the spent fuel pool makeup system.*

7.3   *Order licensees to revise their technical specifications to address requirements to have one train of onsite emergency electrical power operable for spent fuel pool makeup and spent fuel pool instrumentation when there is irradiated fuel in the spent fuel pool, regardless of the operational mode of the reactor.*

7.4   *Order licensees to have an installed seismically qualified means to spray water into the spent fuel pools, including an easily accessible connection to supply the water (e.g., using a portable pump or pumper truck) at grade outside the building.*

7.5   *Initiate rulemaking or licensing activities or both to require the actions related to the spent fuel pool described in detailed recommendations 7.1–7.4.*

4.2.5   *Onsite Emergency Actions*

BACKGROUND

A number of guidelines and procedures guide onsite emergency actions by reactor operators, depending on the nature and extent of events at the plant. As discussed in previous sections, nuclear reactors are designed to handle the loss of offsite electrical power with multiple onsite emergency diesel generators. Events such as a loss of offsite power are within the plants' design basis and are addressed by plant procedures (typically abnormal operating procedures, alarm response procedures, and EOPs). These procedures instruct the plant operators on the steps necessary to take the plant from full-power operation to a safe shutdown condition.

EOPs have long been part of the NRC's safety requirements. The NRC regulations address them through the quality assurance requirements of Criterion V, "Instructions, Procedures, and Drawings," and Criterion VI, "Document Control," in Appendix B to 10 CFR Part 50, and in the administrative controls section of the technical specifications for each plant. Numerous RGs and technical reports (e.g., NUREG-0660, "NRC Action Plan Developed as a Result of the TMI-2 Accident," issued May 1980; NUREG-0737, "Clarification of TMI Action Plan Requirements," issued November 1980; and NUREG-0711, "Human Factors Engineering Program Review Model," issued February 2004) also address EOPs. In addition, the EOPs are the subject of a national consensus standard (American National Standards Institute/ American Nuclear Society 3.21994, "Administrative Controls and Quality Assurance for the Operational Phase of Nuclear Power Plants"). The training and both the written and simulator exams for licensing reactor operators and senior reactor operators also include the EOPs. While implementing EOPs, the event command and control functions remain in the control room under the direction of the shift supervisor and plant manager, both of whom have senior reactor operator licenses.

An SBO is considered to be beyond the plant's design basis. This means that the regulatory requirements stated above do not apply. Instead, specialized requirements included within 10 CFR 50.63 cover SBO. In addition to addressing the loss of offsite electrical power, 10 CFR 50.63 requires reactor licensees to address the simultaneous loss of onsite and offsite electrical power (i.e., SBO) by providing an additional source of electrical power (referred to as an "alternate ac" source) or by showing that the plant could cope with a complete loss of ac power by removing decay heat for a specified period of time. In the case of an SBO, the operators would follow a set of procedures (usually abnormal operating procedures) required by 10 CFR 50.63(c)(ii) and (iii). These procedures would instruct the operators in maintaining safety functions using the alternate ac power source or through coping strategies. In addition, procedures would direct operators to take steps to restore the onsite and offsite sources of ac power.

In addition, the nuclear industry developed SAMGs during the 1980s and 1990s in response to the TMI accident and followup activities. These followup activities included extensive research and study (including several PRAs) on severe accidents and severe accident phenomena. The SAMGs are intended for use by plant technical support staff, usually located in the plant's Technical Support Center (TSC), and are meant to enhance the ability of the operators to manage accident sequences that progress beyond the point where EOPs and other plant procedures are applicable and useful. EOPs typically cover accidents to the point of loss of core cooling and initiation of inadequate core cooling (e.g., core exit temperatures in PWRs greater than 649 degrees Celsius (1,200 degrees Fahrenheit)). As stated in the Westinghouse SAMG documentation, "the SAMG is designed to fill the void between the EOPs and the E-Plan [the procedure and guidance for emergency response]." While implementing SAMGs, the accident command and control functions shift to the TSC and typically to the emergency coordinator or shift technical advisor or both.

In GL 88-20, Supplement 2, "Accident Management Strategies for Consideration in the Individual Plant Examination Process," dated April 4, 1990, the NRC encouraged but did not require licensees to develop and implement SAMGs. Since the SAMGs are voluntary and targeted to technical support staff, the formal training and licensing of plant operators does not address them.

Following the terrorist events of September 11, 2001, the NRC issued security advisories, orders, license conditions, and ultimately a new regulation (10 CFR 50.54(hh)) to require licensees to develop and implement guidance and strategies intended to maintain or restore core cooling and containment and spent fuel pool cooling capabilities under the circumstances associated with the loss of large areas of the plant due to fire or explosion. These requirements have led to the development of EDMGs at all U.S. nuclear power plants. The guidelines and strategies included in the EDMGs are NRC requirements, and the NRC inspected EDMG implementation following the events at Fukushima under TI 2515/183, "Followup to the Fukushima Daiichi Nuclear Station Fuel Damage Event," dated March 23, 2011. The results of this TI are available on the NRC Web site at http://www.nrc.gov/NRR/OVERSIGHT/ASSESS/follow-up-rpts.html. In addition, the NRC added the requirements of 10 CFR 50.54(hh) to the agency's routine inspection program as part of the triennial fire protection inspections.

As stated in the industry's guidance document NEI 06-12, "B.5.b Phase 2&3 Submittal Guideline," Revision 2, issued December 2006 (Agencywide Documents Access and Management System (ADAMS) Accession No. ML070090060), "the initial EDMGs are not a type of emergency operating procedures (EOPs), nor are they intended to be a replacement for EOPs. They are, in fact, intended to be used when the normal command and control structure is disabled and the use of EOPs is not feasible." In terms of command and control, either control room, plant, TSC, or emergency operations facility (EOF) staff could make EDMG decisions. The EDMGs do not play a large role in the formal training and licensing of plant operators.

TASK FORCE EVALUATION

The accidents at Fukushima highlight the importance of having plant operators who are well prepared and well supported by technically sound and practical procedures, guidelines, and strategies. In addition, it is clear that a preplanned approach to command and control and decisionmaking during an emergency is vital.

Each of the onsite emergency action programs (the abnormal operating procedures, alarm response procedures, EOPs, SAMGs, and EDMGs) contributes to overall emergency response capability of plant and operators to mitigate accidents. It is clear that the SAMGs and EDMGs complement the EOPs in an important way. The NRC and industry have established the command and control responsibilities for each of these programs, although not necessarily in a consistent manner. Each of these programs was developed at a different time to serve a different purpose, and each of these programs is treated differently in the NRC's regulations, inspection program, and licensing process, as well as in licensee programs and organizations.

Soon after its establishment, the Task Force considered whether to include the SAMGs as a regulatory requirement and whether to require the integration of the SAMGs into a coherent and holistic program for onsite emergency response activities. This was based on the important role the SAMGs would play in onsite emergency response—just as important as that of EOPs and EDMGs, both required programs. To gain insights into the current implementation of the SAMGs, the Task Force requested that NRC inspectors collect information on how each licensee had implemented that industry voluntary initiative. The inspectors collected information on the initial implementation, ongoing training, and maintenance of the SAMGs under TI 2515/184. The results of the inspection under the SAMG TI reinforced the value of making SAMGs a requirement. The inspectors observed inconsistent implementation of SAMGs and attributed it to the voluntary nature of this initiative. The results of this TI are available on the NRC Web site at http://www.nrc.gov/NRR/OVERSIGHT/ASSESS/SAMGs.html.

In order to ensure the effectiveness of these onsite emergency action programs and to support the effectiveness of mitigation, sound training as well as other readiness measures are important. The inspections under TI 2515/184 identified that some licensee programs include extensive classroom and simulator training and testing on SAMGs, while others do not.

The Task Force concludes that all U.S. plants have addressed all of the elements of onsite emergency actions that need to be accomplished by reactor operators. However, the overall effectiveness of those programs could be substantially enhanced through further integration, including clarification of transition points, command and control, decisionmaking, and through rigorous training that includes conditions that are as close to real accident conditions as feasible.

The Task Force also concludes that action is warranted to confirm, augment, consolidate, simplify, and strengthen current regulatory and industry programs in a manner that produces a single, comprehensive framework for accident mitigation, built around NRC-approved licensee technical specifications. These modified technical specifications would consolidate EOPs, SAMGs, EDMGs, and other important elements of emergency procedures, guidance, and tools in a manner that would clarify command and control and decisionmaking during accidents.

Integration of these accident support functions in a logical and coherent manner and with appropriate regulatory treatment to ensure the effectiveness of operator actions during events would substantially increase the effectiveness of the overall event mitigation. Since the current requirements in this area apply only to normal operation and emergencies within the plant's design basis, they appear outdated and inconsistent with Commission decisions in policy statements and rulemakings to regulate accident mitigation in other areas beyond the plant's design basis. The Task Force concludes that an expansion of the regulatory requirements to include procedures for beyond-design-basis events is warranted, and that such an expansion would redefine the scope of such activities to include them in the regulatory framework to provide defense-in-depth and to ensure adequate protection of public health and safety.

The new treatment of accident procedures would also include the authority to implement SAMGs and EDMGs as necessary and as described or referenced in the plant technical specifications without the need to seek NRC permission or to invoke 10 CFR 50.54(x) and (y). This change would further clarify authority, streamline decisionmaking, and prevent potential delays in taking important emergency actions.

The effectiveness of onsite emergency actions is a very important part of the overall safety of nuclear power plants. The NRC could strengthen the current system substantially by requiring more formal, rigorous, and frequent training of reactor operators and other onsite emergency response staff on realistic accident scenarios with realistic conditions.

Recommendation 8

The Task Force recommends strengthening and integrating onsite emergency response capabilities such as EOPs, SAMGs, and EDMGs.

The Task Force recommends that the Commission direct the staff to further enhance the current capabilities for onsite emergency actions in the following ways:

8.1 *Order licensees to modify the EOP technical guidelines (required by Supplement 1, "Requirements for Emergency Response Capability," to NUREG-0737, issued January 1983 (GL 82-33), to (1) include EOPs, SAMGs, and EDMGs in an integrated manner, (2) specify clear command and control strategies for their implementation, and (3) stipulate appropriate qualification and training for those who make decisions during emergencies.*

- The Task Force strongly advises that the NRC encourage plant owners groups to undertake this activity rather than have each licensee develop its own approach. In addition, the Task Force encourages the use of the established NRC practice of publishing RG (rather than NUREGs, supplements to NUREGs, or GLs) for endorsing any acceptable approaches submitted by the industry.

8.2 *Modify Section 5.0, "Administrative Controls," of the Standard Technical Specifications for each operating reactor design to reference the approved EOP technical guidelines for that plant design.*

8.3 *Order licensees to modify each plant's technical specifications to conform to the above changes.*

8.4 *Initiate rulemaking to require more realistic, hands-on training and exercises on SAMGs and EDMGs for all staff expected to implement the strategies and those licensee staff expected to make decisions during emergencies, including emergency coordinators and emergency directors.*

4.3 EMERGENCY PREPAREDNESS

If mitigation is not successful in preventing a release of radioactive materials from the plant, EP ensures that adequate protective actions are in place to protect public health and safety. Protective actions are taken to avoid or reduce radiation dose. As a condition of their license, operators of nuclear power plants must develop and maintain EP plans that meet comprehensive NRC EP requirements.

Following the TMI accident in 1979, the NRC revised its regulations to substantially enhance EP requirements. The regulations in 10 CFR 50.47, "Emergency Plans," include the 16 planning standards of 10 CFR 50.47(b), and Appendix E, "Emergency Planning and Preparedness for Production and Utilization Facilities," to 10 CFR Part 50 describes information needed to demonstrate compliance with EP requirements. The NRC considers the evaluation criteria in NUREG-0654, "Criteria for Preparation and Evaluation of Radiological Emergency Response Plans and Preparedness in Support of Nuclear Power Plants," Revision 1, issued November 1980, to be guidance and an acceptable means for demonstrating compliance with the Commission's regulations.

After the events of September 11, 2001, the NRC staff reviewed the EP basis for nuclear power plants, considering the impact of hostile actions unanticipated at the time the basis was established. The staff concluded that the EP basis remains valid. The staff based its conclusion on studies that confirmed that the timing and magnitude of releases related to hostile actions would be no more severe than those associated with the other accident sequences considered in the EP basis. However, hostile actions could present unique challenges to EP programs since they differ from operational events for which licensees and offsite response organizations typically plan, train, and exercise. The accident at Fukushima provides its own unique challenges to EP; however, they are rooted in more traditional accident progression sequences.

The agency is in the final stages of a comprehensive rulemaking effort to revise EP regulations, as outlined in SECY-11-0053, "Final Rule: Enhancements to Emergency Preparedness Regulations (10 CFR Part 50 and 10 CFR Part 52)," dated April 8, 2011. These revisions codify the hostile-action-based enhancements, among others. The Task Force has reviewed the draft final rule, particularly the provisions that provide enhancements that would also address challenges to EP programs caused by an accident like that at Fukushima. The subsections below discuss these provisions.

In analyzing the accident and its impact on EP in the United States, the Task Force focused primarily on EP considerations for multiunit and prolonged SBO events, as discussed in the following section. Section 4.3.2 captures other EP insights beyond that focus.

4.3.1 Emergency Preparedness Considerations for Multiunit Events and Prolonged Station Blackout

The accident at Fukushima has shown that prolonged SBO and multiunit events are realities that must be addressed as part of EP. While of low probability, these events have the potential for severe consequences that require an effective EP response. The Task Force's evaluation in this section focuses on a licensee's capability to respond during these types of events.

Currently, the United States has 29 single-unit sites, 33 dual-unit sites, and 3 triple-unit sites. The agency is currently reviewing new reactor applications that may add units to existing sites; however, no applicant has requested to bring the total number of units at a single site to more than four. In most cases, proposed quadruple-unit sites have physical separation between the two existing and the two proposed units.

BACKGROUND

Requirements for emergency response personnel, staffing, and emergency worker protection are codified in 10 CFR 50.47(b)(1), 10 CFR 50.47(b)(2), and 10 CFR 50.47(b)(11).

The proposed EP rulemaking amends Appendix E to 10 CFR Part 50 to address concerns about the assignment of tasks or responsibilities to on-shift emergency response organization (ERO) personnel that would potentially overburden them and prevent the timely performance of their functions under the emergency plan. Licensees must have enough on-shift staff to perform specified tasks in various functional areas of emergency response 24 hours a day, 7 days a week.

The regulation in 10 CFR 50.47(b)(2) requires adequate on-shift staffing levels but gives no clear definition of "adequate." This provides some flexibility in how licensees assign emergency plan implementation duties to on-shift personnel. The proposed EP rule better ensures sufficient on-shift staffing in the threat environment after September 11, 2001, by limiting the assignment of responsibilities that on-shift ERO members would likely perform concurrently with their emergency plan functions.

The regulations in 10 CFR 50.47(b)(1) and 10 CFR 50.47(b)(2) require the assignment and definition of emergency response responsibilities to address decisionmaking and command and control.

Dose assessment, or dose projection, as required by 10 CFR 50.47(b)(9), is the primary means for assessing the potential consequences of a radiological emergency. Appendix E to 10 CFR Part 50 describes the required content of emergency plans, including assessment of the impact of the release of radioactive materials.

After declaration of an emergency, licensees must notify the NRC and State and local response organizations. In addition, licensees are responsible for providing notification and instruction to the public within the plume exposure pathway emergency planning zone (EPZ), as required in 10 CFR 50.27(b)(5) and Section IV.D.3 of Appendix E to 10 CFR Part 50. The predominant method used in the United States for alerting the public is an alert and notification system based on sirens to provide an acoustic warning signal. Some sites employ other means, such as tone alert radios and route alerting, as either primary or supplemental alerting methods. The public then typically receives information about an event and offsite protective actions via emergency alert system broadcasts or other means, such as mobile loudspeakers. The State and local governments are responsible for activating systems to alert and notify the public, and FEMA evaluates this capability.

The regulations in 10 CFR 50.47(b)(6) and 10 CFR 50.47(b)(8) cover the communications equipment needed during an emergency.

As a result of the TMI accident, the NRC and others recognized a need to substantially improve the NRC's ability to acquire data on plant conditions during emergencies. The Commission has defined the NRC's role in the event of an emergency primarily as one of monitoring the licensee to ensure that appropriate recommendations are made with respect to offsite protective actions. Other aspects of the NRC role include supporting the licensee with technical analysis and logistic support, supporting offsite authorities (including confirming the licensee's recommendations to offsite authorities), keeping other Federal agencies and entities informed of the status of the incident, and keeping the media informed based on the NRC's knowledge of the status of the incident, including coordination with other public affairs groups.

To fulfill its role, the NRC requires accurate and timely data on four types of parameters: (1) core and coolant system conditions must be known well enough to assess the extent or likelihood of core damage, (2) conditions inside the containment building must be known well enough to assess the likelihood and consequence of its failure, (3) radioactivity release rates must be available promptly to assess the immediacy and degree of public danger, and (4) data from the plant's meteorological tower are necessary to assess the likely patterns of potential or actual impact on the public.

Experience with the voice-only emergency communications link, previously used for data transmission, has demonstrated that excessive amounts of time are needed for the routine transmission of data and for verification or correction of data that appear questionable. Therefore, the NRC selected the Emergency Response Data System (ERDS) to fulfill the agency's data collection needs. ERDS allows for the direct electronic transmission of selected parameters from the electronic data systems that are currently installed at licensee facilities. ERDS was designed for use only during emergencies and would be activated by the licensees during declared emergencies classified at the Alert lever or higher to begin transmission to the NRC Operations Center. ERDS would be supplemented with voice transmission for essential data not available on the licensee's systems, rather than require a modification to the existing system to transmit that data.

Section IV of Appendix E to 10 CFR Part 50 codifies the requirements for ERDS. The NRC requires power reactor licensees to transmit ERDS data to a server at NRC Headquarters. Many licensees currently use analog modulator/demodulators (modems) to establish point-to-point data connections. Although this technology was state of the art when ERDS was first implemented, it is now obsolete, and replacement equipment is no longer readily available. In addition, the use of modems inherently introduces a cyber security vulnerability to the systems to which the modems are attached.

As part of the current effort to modernize the ERDS infrastructure, the NRC has been working with individual licensees to develop an acceptable solution to replace the existing modems. The NRC chose virtual private network (VPN) technology to create a secure point-to-point data pathway between the licensee site and NRC Headquarters. This technology permits all ERDS-enabled facilities to connect to the NRC simultaneously, thereby enhancing the NRC's ability to respond to incidents that may affect multiple licensees simultaneously, such as grid instability events. Some licensees are currently in the process of replacing the existing analog modems with VPN devices on a voluntary basis, while others have yet to commit to the initiative.

The regulation in 10 CFR 50.47(b)(14) requires periodic drills and exercises to develop key skills and evaluate major portions of emergency response capabilities, and 10 CFR 50.47(b)(15) requires radiological emergency response training. NUREG-0654 includes further detail on the implementation of these regulations.

The regulation in 10 CFR 50.47(b)(8) requires the provision and maintenance of adequate emergency facilities and equipment to support the emergency response. Studies of the TMI accident identified the need for extensive improvements in the overall response to accidents at nuclear plants, including enhanced facilities and systems to support the control room in mitigating the consequences of accidents and to support the licensee's capability to respond to abnormal plant conditions. NUREG-0696, "Functional Criteria for Emergency Response Facilities," issued February 1981, describes the facilities and systems that licensees can use to improve emergency response to accidents, such as the TSC, operational support center (OSC), and EOF. The document also provides guidance on the functional criteria for these facilities and on the integrated support these facilities offer to the control room.

Licensees must have the capability to augment the on-shift staff within a short time after the declaration of an emergency. To accomplish this, licensees typically staff an onsite TSC, which relieves the control room of emergency response duties and allows operators to focus on reactor plant safety. Responders also staff an onsite OSC to provide an assembly area for damage repair teams. Finally, licensees establish an EOF to function as the center for evaluation and coordination activities related to the emergency and as the focal point for providing information to Federal, State, Tribal, and local authorities involved in the response.

The proposed EP rulemaking amends Section IV.E.8 of Appendix E to 10 CFR Part 50 to address concerns about ERO augmentation during hostile action by requiring licensees to identify alternative facilities as staging areas for responding staff. This alternate site must have the capability for communication with the EOF, control room, and plant security; the capability to perform offsite notifications; and the capability for engineering assessment activities, including damage control team planning and preparation.

Licensees have submitted several requests to the NRC to consolidate EOFs for plants they operate within a State or in multiple States. Previous regulatory standards did not address the capabilities and functional requirements for a consolidated EOF, such as capabilities for handling simultaneous events at two or more sites. The NRC is revising, via the proposed EP rule, the regulations and associated guidance to reflect a performance-based approach for EOFs and to provide functional requirements for these facilities.

TASK FORCE EVALUATION

The events in Japan have highlighted the importance of the human element during response to an emergency. The current regulatory approach provides for staffing and protection of staff during postulated reactor accidents. External events on the scale of Fukushima cause concerns that the existing framework would present challenges to personnel and their safety. In particular, the presence of staff onsite during the initiation of the emergency condition, staff needed to augment the current onsite staff, and staff needed for crew relief poses significant challenges during large external events.

The proposed EP rulemaking will require an analysis of the duties of on-shift personnel to ensure the fulfillment of all required functions. The accident at Fukushima supports an expansion of this analysis to include whether enough responders are available for multiunit events. During traditional accident scenarios, most required response personnel would be summoned to the site within a certain required time. During a catastrophic natural disaster, the local infrastructure may challenge the timely augmentation of on-shift personnel. Should licensees choose not to add more on-shift personnel, they would need to have a viable notification and transportation strategy for ensuring that the staff needed to augment the site response would be available to respond effectively. In addition, since more responders are needed during a multiunit event, additional protective equipment would be needed.

Through discussions with inspectors and technical experts, the Task Force has found that the current framework for command and control during an accident has been developed and practiced over the years and that authorities for decisionmaking during an event are well defined. However, in this area as well, multiunit events create a nuance to the command and control structure that is not yet fully developed.

During an accident like that at Fukushima, it may be necessary to make difficult decisions to prioritize limited response resources. The EOF remains the primary facility for interaction with offsite authorities, and the TSC remains the primary facility for technical response to the accident. Currently, during a General Emergency, the licensee's emergency director assigned with the authority to lead the licensee response is located in the EOF. An emergency director in the TSC remains in command of the technical assessment and damage control aspects of the response. During a multiunit event, the lead TSC official would be in the best position to address the triage and prioritization of resource requirements for each unit.

Ensuring that the response framework contains the correct level of authority, knowledge, and experience is paramount to successful response. In light of the Fukushima accident, the staff should explore concepts such as whether decisionmaking authority is in the correct location (i.e., at the facility), whether currently licensed operators need to be integral to the ERO outside of the control room (i.e., in the TSC), and whether licensee emergency directors should have a formal "license" qualification for severe accident management in addition to their existing qualification requirements, and different than a reactor operator license.

The accident at Fukushima also presented challenges with respect to dose assessment capability because of the multiunit nature of the release. While monitoring the accident, dose assessors at the NRC had to use makeshift, ad hoc methods to consider the source term from these multiple concurrent releases and overlay the release points to arrive at a final sitewide dose projection. The difficulty of conducting this dose assessment highlights the gap in capability for U.S. plants to perform multiunit dose assessment. Currently, dose assessment software (such as the NRC's RASCAL) is not designed to model multiunit accidents. The presence of releases from multiple units and spent fuel pools at Fukushima has highlighted the need for the ability to project doses from releases at multiple units.

In the absence of a software solution, licensees would need to develop emergency plan guidance that outlines how to conduct a multiunit dose assessment, including spent fuel pools as release points, to ensure that this capability exists.

Following a large natural disaster, such as the accident at Fukushima, the ability to notify both government authorities and the public could be challenged. Licensees use traditional telephones, cellular telephones, satellite telephones, short wave radio, and the Internet to communicate with the NRC and State and local governments. The Task Force believes that licensees have enough redundant and diverse methods to communicate with the NRC and State and local governments that it is reasonable to expect the successful communication of a declared emergency onsite. Because of the diverse nature of these methods, it is unlikely that a common-cause failure would disable all means of communication.

With respect to alerting the public within the plume exposure pathway EPZ, the Task Force finds that the provisions being incorporated by the EP rulemaking will enhance the existing public alerting framework. The proposed final rule addresses provisions for a backup to the primary alert and notification system. The backup measures would be implemented if the primary means of alert and notification were unavailable during an emergency. These enhancements will provide an additional layer of preparedness that will be useful during a large-scale natural disaster.

Communications equipment plays a critical role in the effective conduct of any incident response effort. Currently, licensees use an array of different telecommunications devices to communicate onsite, including hardwired telephones, cellular telephones, satellite telephones, radios, and pagers. Since many of these devices depend on electrical power, most by battery, their use during a prolonged SBO would be limited by battery life. Additionally, since hardwired telephones and cellular telephones rely upon offsite infrastructure (i.e., phone switches and cell towers) that could be damaged or destroyed by the event, their use may be limited. Onsite radios may be hampered by a lack of power to radio repeaters.

During the accident at Fukushima, numerous organizations assessed the evolution of the situation onsite and sought data to aid decisionmakers during the response. Having accurate, real-time data from the site allows for the performance of a multitude of analyses. In addition, the more data that can be provided using automated sources, the less burden placed on the licensee to provide information. Having data provided directly from automated sources at the site also gives confidence to government authorities and the public that the plant operator is not filtering the details of an evolving accident. However, the current regulatory approach and requirements do not ensure that ERDS data would be available during a prolonged SBO or during other natural disasters when power supplies could be lost and transmission capability may be affected.

Another challenge evident from Fukushima is the need for archived data to aid in the reconstruction of events after an accident. When ERDS data are transmitted to the NRC servers, the NRC stores the incoming data. Ensuring that licensees have reliable power sources and transmission methods will therefore ensure that data are archived for accident reconstruction.

Effective onsite and offsite response necessitates abundant communication. During a prolonged SBO, an awareness of plant conditions becomes even more critical. Ensuring that the NRC and State stakeholders continue to receive plant data, especially during the most severe accidents, adds another layer of both technical assessment capability and confidence that onsite conditions are being communicated in a timely and accurate way.

Effective and timely ERDS modernization is necessary to ensure that during multiunit events, the NRC is able to receive data from all affected nuclear units. Given the current limited modem connections available for simultaneous use, the NRC should ensure that licensees upgrade to the VPN solution in an expedient manner. Once the current modernization effort is complete, the staff should then evaluate additional modernization to address redundant transmission capability, completeness of data, and continuous monitoring.

The conduct of training and exercises provides confidence that emergency plans are workable and that personnel would be successful in mitigating the consequences of an accident. However, EP drills and exercises currently do not consider prolonged SBO or multiunit accident scenarios. Therefore, training and exercises should explore and practice the concepts of command and control, decisionmaking, prioritization, and contingency planning under these conditions. Since licensees will likely rely heavily on offsite support during this type of event, exercising the steps necessary to identify and acquire offsite equipment and support is an important element to practice.

Both prolonged SBO and multiunit events present new challenges to EP facilities that were not considered when the NRC issued NUREG-0696. The accident at Fukushima has clearly shown that these events are a reality. While several utilities have implemented combined EOFs that are capable of handling multiunit events, licensee onsite TSCs and OSCs have not been designed or drilled for multiunit events. TSCs and OSCs are not as conducive to consolidation since most job functions in these facilities would be specific to a particular unit, and attempts to have one person manage the technical response for multiple units could introduce opportunities for errors. The proposed EP rulemaking requires the identification of alternate facilities. During a multiunit event, licensees could use both the normal and alternate facilities to provide enough capacity for response. The proposed EP rulemaking codifies performance-based requirements for combined EOFs. EOFs are more suited to handling multiunit events since they represent the interface with offsite responders, regardless of the number of onsite units in emergency status. The NRC should use a performance-based approach to TSCs, OSCs, and EOFs to ensure flexibility for licensees.

Recommendation 9

The Task Force recommends that the NRC require that facility emergency plans address prolonged SBO and multiunit events.

The Task Force recommends that the Commission direct the staff to do the following:

9.1 *Initiate rulemaking to require EP enhancements for multiunit events in the following areas:*

- personnel and staffing
- dose assessment capability
- training and exercises
- equipment and facilities

9.2   *Initiate rulemaking to require EP enhancements for prolonged SBO in the following areas:*

- communications capability
- ERDS capability
- training and exercises
- equipment and facilities

9.3   *Order licensees to do the following until rulemaking is complete:*

- Determine and implement the required staff to fill all necessary positions for responding to a multiunit event.

- Add guidance to the emergency plan that documents how to perform a multiunit dose assessment (including releases from spent fuel pools) using the licensee's site-specific dose assessment software and approach.

- Conduct periodic training and exercises for multiunit and prolonged SBO scenarios. Practice (simulate) the identification and acquisition of offsite resources, to the extent possible.

- Ensure that EP equipment and facilities are sufficient for dealing with multiunit and prolonged SBO scenarios.

- Provide a means to power communications equipment needed to communicate onsite (e.g., radios for response teams and between facilities) and offsite (e.g., cellular telephones, satellite telephones) during a prolonged SBO.

- Maintain ERDS capability throughout the accident.

9.4   *Order licensees to complete the ERDS modernization initiative by June 2012 to ensure multiunit site monitoring capability.*

Recommendation 10

The Task Force recommends, as part of the longer term review, that the NRC should pursue additional EP topics related to multiunit events and prolonged SBO.

The Task Force recommends that the Commission direct the staff to do the following:

10.1 *Analyze current protective equipment requirements for emergency responders and guidance based upon insights from the accident at Fukushima.*

10.2 *Evaluate the command and control structure and the qualifications of decisionmakers to ensure that the proper level of authority and oversight exists in the correct facility for a long-term SBO or multiunit accident or both.*

- Concepts such as whether decisionmaking authority is in the correct location (i.e., at the facility), whether currently licensed operators need to be integral to the ERO outside of the control room (i.e., in the TSC), and whether licensee emergency directors should have a formal "license" qualification for severe accident management.

10.3 *Evaluate ERDS to do the following:*

- Determine an alternate method (e.g., via satellite) to transmit ERDS data that does not rely on hardwired infrastructure that could be unavailable during a severe natural disaster.

- Determine whether the data set currently being received from each site is sufficient for modern assessment needs.

- Determine whether ERDS should be required to transmit continuously so that no operator action is needed during an emergency.

4.3.2 Other Emergency Preparedness Insights

In addition to prolonged SBO and multiunit events, the Task Force has identified other EP insights from the accident at Fukushima. An overarching lesson is that major damage to infrastructure in the area surrounding the plant might challenge an effective emergency response. In addition, the NRC should learn from the real-world implementation of protective actions and further develop concepts such as recovery and reentry. A third lesson relates to the need for the public to be educated on radioactivity and radiological hazards before an incident occurs.

BACKGROUND

Natural disasters, such as hurricanes, floods, large fires, and tornados, occur often in the United States. State and local governments constantly manage these types of events and ensure that emergency plans are workable. Likewise, the NRC works closely with FEMA to ensure that the state of offsite preparedness continues to ensure reasonable assurance that the public can be protected should an accident occur. While these disasters do not significantly impact nuclear plants because of the plants' robust design features, they often cause significant damage in the surrounding community and could challenge the effective implementation of an offsite emergency plan should it be required. These natural disasters may alter the preplanned emergency framework by changing evacuation routes (e.g., bridge washed out or tree down in the roadway), disabling emergency sirens, or making evacuation dangerous (e.g., icy roads). Local government decisionmakers take factors such as these into consideration when formulating protective action decisions to determine how best to protect the public. A catastrophic natural disaster would create additional challenges beyond those routinely experienced by state and local emergency planners.

Currently, if a nuclear power plant shuts down as a result of a natural disaster and the disaster is of such severity that damage or changes to the offsite emergency response infrastructure may be substantial or are in question, the NRC and FEMA determine the status of offsite EP and coordinate approval of plant restart activities.

Evacuation time estimates (ETEs) are used as a tool to (1) develop and improve evacuation plans in advance of an accident and (2) to decide whether sheltering or evacuation is the more protective response during an accident. Evacuation is preferred if a dose in excess of protective action guides is probable, but it is not always more effective in reducing public exposure. ETEs are a planning tool and do nothing to affect conditions during an actual evacuation. Draft NUREG/CR-7002, "Criteria for Development of Evacuation Time Estimate Studies," issued May 2010, provides the latest guidance for licensees to develop a comprehensive set of ETEs.

Infrastructure damage could also affect the ability of licensees to bring equipment from offsite sources to the site. The regulation in 10 CFR 50.47(b)(3) requires licensees to maintain agreements with offsite organizations needed to support emergency response. Typically, this includes local fire departments, law enforcement, and medical support. As part of the requirements of 10 CFR 50.54(hh)(2)(i), licensees must identify and seek readily available agreements for additional offsite resources (both local and regional) that could support fire fighting, electrical power, core cooling, and other needs.

The regulation in 10 CFR 50.47(b)(10) requires licensees to make protective action recommendations. To facilitate a preplanned strategy for protective actions during an emergency, two EPZs surround each nuclear power plant. The exact size and shape of each EPZ is determined through detailed planning that includes consideration of the specific conditions at each site, unique geographical features of the area, and demographic information. This preplanned strategy for an EPZ provides a substantial basis to support activity beyond the planning zone in the extremely unlikely event it would be needed. The plume exposure pathway EPZ has a radius of about 16 kilometers (10 miles) from the reactor site. Predetermined protective action plans in place for this EPZ are designed to avoid or reduce dose from potential exposure to radioactive materials. These action plans include sheltering, evacuation, and the use of potassium iodide (KI) where appropriate. The ingestion exposure pathway EPZ has a radius of about 80 kilometers (50 miles) from the reactor site. Predetermined protective action plans in place for this EPZ are designed to avoid or reduce dose from the potential ingestion of radioactive materials. These action plans include a ban of contaminated food and water.

NUREG-0396, "Planning Basis for the Development of State and Local Government Radiological Emergency Response Plans in Support of Light Water Nuclear Power Plants," issued December 1978, establishes the concept and basis for the 10-mile and 50-mile EPZs, describes the basis for the types of reactor accidents that should be considered in an emergency plan, states that emergency planning is not based on probability but on public perceptions of the problem and what could be done to protect public health and safety (as a matter of prudence rather than necessity), and concludes that the objective of an emergency plan should be to provide dose savings for a spectrum of accidents that could produce offsite doses in excess of the U.S. Environmental Protection Agency's (EPA's) protective action guides.

The regulations in 10 CFR 50.47(b)(13) also address general plans for recovery and reentry and the means by which decisions to relax protective measures (e.g., allow reentry into an evacuated area) are reached. While requirements for the licensees, States, and local governments appear in 10 CFR 50.47(b)(13) and NUREG-0654, the approach of the U.S. Government to recovery and reentry involves numerous other Federal partners.

Appendix E to 10 CFR Part 50 and 10 CFR 50.47(b)(9) require radiation monitoring, which refers to the measurement of actual radiation dose. NUREG-0654 provides guidance to licensees and offsite response organizations on topics including field monitoring, monitoring equipment, and airborne plume tracking.

In January 2001, the Commission published a rule change to the NRC EP regulations to include consideration of the use of KI. If taken properly, KI may help reduce the dose of radiation to the thyroid gland from radioactive iodine and thus reduce the risk of thyroid cancer. If a person takes in radioactive iodine after consuming KI, the radioactive iodine will be rapidly excreted from the body. The NRC has supplied KI tablets to States requesting it for the population within the 10-mile EPZ. If warranted, KI is to be used to supplement evacuation or sheltering, not to take the place of these actions.

The population closest to the nuclear power plant that is within the 10-mile EPZ is at greatest risk of exposure to radiation and radioactive materials. When the population is evacuated from the area and potentially contaminated foodstuffs are removed from the market, the risk from further radioactive iodine exposure to the thyroid gland is essentially eliminated. Beyond the 10-mile EPZ, the major risk of radioiodine exposure is from the ingestion of contaminated foodstuffs, particularly milk products. Both EPA and the Food and Drug Administration have published guidance to protect consumers from contaminated foods. These protective actions are preplanned in the 50-mile ingestion pathway EPZ.

In the unlikely event of a nuclear power plant accident, it is important to follow the direction of State or local governments in order to make sure that protective actions, such as taking KI pills, are implemented safely and effectively for the affected population. While use of KI at the correct time and dosage can help prevent the uptake of radioactive iodine in the thyroid gland, it does not offer any protection from other radioisotopes or for other organs. KI is a medical tool for a limited, targeted situation.

TASK FORCE EVALUATION

The current regulatory approach for the evaluation of offsite EP following a natural disaster is robust and has proven its effectiveness following recent hurricanes, including Hurricane Katrina. An NRC task force examined the lessons learned from the active 2005 hurricane season in a report dated March 30, 2006 (ADAMS Accession No. ML060900005).

ETEs are currently recalculated when the population around a nuclear plant either increases or decreases significantly. As supported by the proposed EP rule, the scenarios described in NUREG/CR-7002 provide a basis for licensees to develop a comprehensive set of ETEs. Performing additional time estimates for natural disasters with unpredictable damage would offer no corresponding benefit to licensee personnel in providing appropriate protective action recommendations to offsite officials or to offsite emergency planners in developing evacuation and other protective action strategies. With regard to seismic events impacting a plant site and causing a severe accident, consideration of the estimated time to evacuate the area would require assessment of the overall condition of dwellings, buildings, roadways, and other critical infrastructure at the time of the event. Additionally, when dwellings and other buildings are compromised and can no longer provide effective sheltering, evacuation would be conducted regardless of the time needed. If sheltering is not an option, ETE values are not relevant to decisionmaking. With regard to extreme weather conditions such as significant hurricanes, evacuation might involve an area much larger than a nuclear power plant site and begins days in advance of landfall. Therefore, the evacuation of the population in the plant vicinity would likely be completed well before the hurricane affects that area.

The accident at Fukushima has illustrated the potential increased need for offsite assistance to the licensee. In the case of large natural disasters such as earthquakes, hurricanes, and floods, the phenomena challenging the plant will also have affected the local community. In these cases, prearranged resources may not be available because of their inability to reach the plant site, other (potentially lifesaving) priorities within the community, or the destruction of those resources.

During the emergency at Fukushima, conditions deteriorated such that Japanese officials required additional protective actions up to and beyond a 20-kilometer (16-mile) area around Fukushima (i.e., beyond the equivalent of the U.S. plume exposure pathway EPZ). The possibility of making protective action recommendations for areas beyond the plume exposure pathway EPZ has been a program consideration since the inception of the EPZ concept in the United States. The emergency planning basis for U.S. plants, as discussed in NUREG-0654, states, "...detailed planning within 10 miles would provide a substantial base for expansion of response efforts in the event that this proved necessary." NUREG-0654 goes on to state that it would be unlikely that any protective actions for the plume exposure pathway would be required beyond the plume exposure pathway EPZ. It further stated that the plume exposure EPZ is of sufficient size for actions within this zone to provide for substantial reduction in early severe health effects (injuries or deaths) in the event of a worst case core melt accident.

While the U.S. EP framework has always noted that the plume exposure pathway EPZ provides a basis for expansion, insights from real-world implementation at Fukushima, including the realities of multiunit events, might further enhance U.S. preparedness for such an event. The Task Force acknowledges that every situation will differ, so detailed preplanning in this area is not plausible. As information and insights emerge about the challenges faced by Japanese officials while implementing protective actions around Fukushima, the NRC and its partners should evaluate those insights to identify enhancements to the decisionmaking framework in the United States.

Similarly, licensees and States are required to have plans for recovery and reentry; however, these plans remain largely conceptual and are rarely practiced. Since recovery and reentry have proven to present challenges at Fukushima, the NRC should continue work in this area to forward the U.S. Government approach.

In the area of radiation monitoring, licensees have numerous fixed radiation monitors located inside the plant on the property of the site. During an emergency, licensee field teams, in addition to Federal, State, and local teams, are dispatched to take real-time radiation readings offsite. While multiunit events do not specifically change radiation monitoring concepts, a long-term SBO could challenge radiation monitors that rely upon an external power supply. As long as field teams are adequately staffed, equipped, and capable of transit given the nature of the natural disaster, field monitoring remains an effective method to acquire radiation data. In additional, real-time radiation monitoring helps to validate dose projections, quantify actual dose, and provide confidence to the public and stakeholders that the conditions around the site are being accurately characterized.

During a radiological release, accurate and timely dose data are critical to validating dose projections and ensuring dose reduction for the public. Having publicly available dose data provides a level of public confidence. The staff should explore the concepts of ac independence, survivability of equipment during a natural hazard, and ability to transmit real-time radiation readings publicly via the Internet.

The concepts of radiation and radiation safety continue to be challenging topics to teach to those not involved in nuclear fields. This is underscored by the invisible nature of radiation and the potential chronic, in addition to acute, health affects that could be attributed to radiation exposure. While the NRC and various other Federal, State, and local government agencies and other medical organizations have conducted education on the uses, benefits, and drawbacks of KI, recent events have shown a continued gap in the public knowledge with respect to KI.

Based on the observed gaps in public awareness following the accident at Fukushima, an effort to increase education and outreach in the vicinity of each nuclear power plant is warranted. Misinformation and hysteria during a nuclear emergency challenge the agency's goal of public confidence. Training should be targeted to the areas surrounding each nuclear power plant in order to reach those who could be affected should an emergency at a nuclear power plant occur. In addition to public participation, the NRC should make extra effort to involve local response personnel, health officials, decisionmakers, media, and local politicians.

Recommendation 11

The Task Force recommends, as part of the longer term review, that the NRC should pursue EP topics related to decisionmaking, radiation monitoring, and public education.

The Task Force recommends that the Commission direct the staff to do the following:

11.1 *Study whether enhanced onsite emergency response resources are necessary to support the effective implementation of the licensees' emergency plans, including the ability to deliver the equipment to the site under conditions involving significant natural events where degradation of offsite infrastructure or competing priorities for response resources could delay or prevent the arrival of offsite aid.*

11.2 *Work with FEMA, States, and other external stakeholders to evaluate insights from the implementation of EP at Fukushima to identify potential enhancements to the U.S. decisionmaking framework, including the concepts of recovery and reentry.*

11.3 *Study the efficacy of real-time radiation monitoring onsite and within the EPZs (including consideration of ac independence and real-time availability on the Internet).*

11.4 *Conduct training, in coordination with the appropriate Federal partners, on radiation, radiation safety, and the appropriate use of KI in the local community around each nuclear power plant.*

5. IMPLICATIONS FOR NRC PROGRAMS

Section 3 of this report presented the most significant Task Force finding and recommendation relative to NRC programs (i.e., the NRC regulatory framework and how it could be strengthened). This section will discuss impacts on the NRC inspection program, the management of NRC records and information, and NRC participation in international activities.

5.1 NRC INSPECTION PROGRAM

BACKGROUND

For many decades, the NRC has had a vigorous reactor inspection and oversight program for operating reactors and vendor inspection program for new plant construction. The ROP is a formal process integrating the NRC's inspection, assessment, and enforcement programs, and it has been in place for the last decade. The ROP evaluates the overall safety performance of operating nuclear power reactors and communicates this information to licensee management, members of the public, and other stakeholders. In addition, since 2004, the NRC has included safety culture in the ROP for operating reactor licensees. Over the past three decades, the safety performance of the nuclear fleet in the United States has improved, as a result in some part of the effectiveness of the NRC's inspection, assessment, and enforcement programs.

Regarding reactor protection and mitigation systems, a fundamental characteristic of the ROP is that inspection activities or "samples" are selected for the relative risk significance of the activity or equipment being examined based on its effect on core damage frequency. Further, the NRC evaluates inspection findings in these areas and uses the significance determination process to determine significance based on risk. The ROP's reliance on risk undervalues the safety benefit of defense-in-depth and consequently reduces the level of NRC resources focused on inspecting defense-in-depth characteristics that contribute to safety.

In addition, the ROP does not consider the industry's voluntary safety enhancements. Consequently, the staff devotes limited inspection effort to voluntary initiatives such as the implementation and adequacy of SAMGs or the implementation of the Groundwater Protection Initiative. If it inspects them at all, the agency would likely evaluate voluntary initiatives through nonrecurring inspections guided by TIs. Finally, the structure of the risk-based inspection program under the ROP focuses on licensee compliance with regulations and requirements and leaves very limited opportunity for inspection staff to evaluate the adequacy of the licensing basis at a given facility.

TASK FORCE EVALUATION

Following the Fukushima accident, the NRC recognized the more general relevance and importance of the SAMGs and EDMGs that had been developed, and later required, following the accident at TMI and the terrorist events of September 11, 2001. Therefore, the NRC issued TI 2515/183 to its resident inspectors to ensure licensee compliance with existing requirements and to collect information on the readiness of these measures for use under various external challenges.

The Task Force has had the benefit of the results of the inspections under TI 2515/183 and concludes that the effort was well planned and worthwhile. In addition, in an April 22, 2011, memorandum from Charles L. Miller to Eric J. Leeds, "Task Force Request Regarding Inspection of Severe Accident Management Guidelines," the Task Force asked that inspectors collect information on the implementation of the voluntary SAMG initiative. In response, the NRC issued TI 2515/184. The Task Force has factored the information from both of these staff efforts into its evaluation and recommendations.

Through these two inspection activities, the Task Force also had the opportunity to compare industry activities under a required program and a similar voluntary initiative (i.e., EDMGs and SAMGs). Both programs had been effectively implemented, including initial program formulation and licensee staff training. Those programs are now 10 to 20 years old, and some licensees have maintained both programs in a manner expected for an important safety activity, including in terms of maintenance, configuration control, training, and retraining. However, some licensees have treated the industry voluntary initiative (the SAMG program) in a significantly less rigorous and formal manner, so much so that the SAMG inspection would have resulted in multiple violations had it been associated with a required program. The results of the SAMG inspection do not indicate, nor does the Task Force conclude that, the SAMGs would not have been effective if needed. However, indications of programmatic weaknesses in the maintenance of the SAMGs are sufficient to recommend strengthening this important activity.

On the basis of its evaluation, the Task Force concludes that enhancements to the inspection program would improve its focus on safety. Since these modifications are all staff activities not involving any backfit considerations, the NRC can pursue them based solely on their value in improving the staff's safety focus.

Recommendation 12

The Task Force recommends that the NRC strengthen regulatory oversight of licensee safety performance (i.e., the ROP) by focusing more attention on defense-in-depth requirements consistent with the recommended defense-in-depth framework.

The Task Force recommends the Commission direct the staff to strengthen the ROP by doing the following:

12.1 Expand the scope of the annual ROP self assessment and biennial ROP realignment to more fully include defense-in-depth considerations.

12.2 Enhance NRC staff training on severe accidents, including training resident inspectors on SAMGs.

5.2   MANAGEMENT OF NRC RECORDS AND INFORMATION

BACKGROUND

In the 1990s, the NRC decided to prepare for a transition from paper records management to electronic records management and began a process that resulted in the deployment of ADAMS in the late 1990s and early 2000s. At present, essentially every document generated by the agency is effectively integrated into a powerful electronic records management system that is searchable in a variety of ways. However, at the time of the original transition to ADAMS, individual offices decided what records generated before the deployment of ADAMS would be incorporated into the electronic system and what records would remain in paper and microfiche format.

In SECY-06-0164, "The NRC Knowledge Management Program," dated July 25, 2006, the staff described to the Commission the plans for developing and implementing an agency knowledge management program. The program recognized that the agency is a knowledge-centric organization that depends on its staff to make sound regulatory decisions to accomplish the agency's mission of protecting people and the environment. The staff informed the Commission about the NRC's ongoing programs for maintaining explicit, implicit, and tacit knowledge and new initiatives to leverage information technology solutions to transfer knowledge. The knowledge management program focused on a variety of areas to enhance identifying, retaining, transferring, and using critical information and knowledge in the agency's ongoing work. The staff updated the Commission on the implementation status of knowledge management initiatives in SECY-07-0138, "NRC Knowledge Management Program Status Update," dated August 14, 2007.

TASK FORCE EVALUATION

During the course of data collection and analysis, the Task Force searched for and reviewed hundreds of documents, including some issued 40 or more years ago. The Task Force also interacted with scores of NRC staff and other individuals to understand issues and evaluate insights from the Fukushima event.

During the course of these activities, the Task Force struggled to access electronically a variety of critical historical documents, including Commission papers and staff requirements memoranda, GLs, bulletins, NUREGs, RGs, and other key documents that significantly contribute to the knowledge base for technical and regulatory evaluation and decisions made by the agency. Many documents dated before 2000 are not available electronically. Consequently, from the perspective of integrated knowledge management, it would be equally difficult for staff to access these foundational documents without personal paper files.

In addition, during a variety of interviews with staff from multiple offices on similar subjects, it became clear that organizationally influenced knowledge gaps may exist, in that staff members in one organization may not be aware of information in another organization that is relevant to their work. In addition, staff members working on operating reactor licensing noted that, in order to support a licensing action or decision, they often need to read, evaluate, and refer to critical documents that must be viewed on microfiche.

The NRC has made great strides in electronic document management. Because of financial constraints, legacy documents are not all accessible electronically through ADAMS, and no strategy exists to further enhance the availability of historical documents underpinning agency positions and decisions. The Task Force believes that the staff should strongly consider continued retrofitting of its electronic document library in a methodical and coherent way to incorporate all key legacy documents.

Further, the Task Force believes that the NRC would enhance its organizational and technical capacity by establishing an electronic framework, similar to Wikipedia and managed by staff technical experts, that links data on various subjects so that the staff members could access the information easily, allowing them to develop the well-founded knowledge required to accomplish their tasks and support their regulatory decisions.

As time passes, maintaining historical institutional knowledge becomes more dependent on a comprehensive and accessible information management system. The Task Force therefore encourages the staff to do the following:

- Add key legacy documents to the electronic records system (ADAMS).

- Develop an information management platform to enhance organizational and technical capacity by providing subject matter in an electronic format managed by staff technical experts.

5.3   INTERNATIONAL COOPERATION AND COORDINATION

BACKGROUND

The NRC is actively involved in numerous international activities related to the Fukushima accident, both bilaterally and multilaterally, through such organizations as IAEA, NEA, and the Multinational Design Evaluation Program (MDEP). The regulatory authority of Japan (the Nuclear and Industrial Safety Agency with its support organization, the Japan Nuclear Energy Safety Organization) is also a member of each of these international organizations. Through each of the venues, the NRC has received information on the accident in Japan and on the followup activities of each nation.

In addition to its frequent contacts through the NRC Operation Center, the NRC staff has participated in the following activities, all of which included discussions of the Fukushima accident and followup actions planned or underway:

- April 2011 NEA Steering Committee meeting

- April 2011 MDEP Steering Technical Committee meeting

- May 2011 MDEP EPR reactor design working group meeting

- May 2011 NEA Committee on Nuclear Regulatory Activities (CNRA) planning meeting

- June 2011 CNRA discussions on Fukushima

- June 2011 NEA, IAEA, France G20-G8 Forum on the Fukushima Accident

- June 2011 IAEA Ministerial Meeting on Fukushima

These international interactions have helped the world's nuclear safety authorities share information and develop plans to enhance safety following the Fukushima accident. In part because of these interactions, the nuclear regulatory authorities around the world have developed a consistent understanding of the important elements of the accident and similar roadmaps forward, each appropriate to the individual country and its regulatory approach. The staff has also been well informed of the activities of the European regulatory authorities in the Western European Nuclear Regulators' Association and the European Nuclear Safety Regulators Group through discussions within MDEP, NEA/CNRA, and IAEA.

TASK FORCE EVALUATION

The NRC's longstanding commitment to international cooperation has benefitted the agency during the monitoring and followup activities related to the Fukushima accident. It is clear that the NRC's active participation in international activities has served and continues to serve nuclear safety in the United States and worldwide. It is also clear that the NRC's long-term activities, such as further information collection, event analysis, and comparisons of the effectiveness of actions taken, would benefit from international cooperation both in terms of effectiveness and cost efficiency.

International cooperation to date has been very effective in collecting and sharing information and in developing and confirming an understanding of the important elements of the Fukushima accident relevant to the activities of the Task Force. NRC participation in these international efforts also contributes to the safety of nuclear reactors throughout the world by providing forums for sharing the agency's best understanding of events and issues and effectively addressing them. These international forums also enable the NRC to provide leadership in the resolution of important safety concerns.

In addition, participation in international efforts generally requires a minimal commitment of resources and in some cases, such as the collection and analysis of accident data, actually results in a resource savings.

The Task Force endorses continued international cooperation and coordination, including the following:

- participation in collaborative, international efforts to determine and analyze the Fukushima accident sequence of events

- participation in international efforts to update IAEA Fundamental Safety Standard (NSR-1) and other related standards to reflect insights from the Fukushima event

- continued cooperation and coordination with other national regulatory authorities on insights from the Fukushima event as well as their plans, actions, and findings

This page intentionally left blank

Enhancing Reactor Safety in the 21st Century

6. SUMMARY OF OVERARCHING RECOMMENDATIONS

This section presents the Task Force's recommendations for improving the safety of both operating and new nuclear reactors. It also addresses recommended improvements in the NRC programs for the oversight of reactor safety. The recommendations are based on the Task Force's evaluations of the relevant issues identified from the Fukushima accident. Appendix A of this report proposes an implementation strategy and offers further details on these recommendations.

The Task Force makes the following overarching recommendations, as stated in the preceding sections of this report:

Clarifying the Regulatory Framework

1. The Task Force recommends establishing a logical, systematic, and coherent regulatory framework for adequate protection that appropriately balances defense-in-depth and risk considerations. (Section 3)

Ensuring Protection

2. The Task Force recommends that the NRC require licensees to reevaluate and upgrade as necessary the design-basis seismic and flooding protection of SSCs for each operating reactor. (Section 4.1.1)

3. The Task Force recommends, as part of the longer term review, that the NRC evaluate potential enhancements to the capability to prevent or mitigate seismically induced fires and floods. (Section 4.1.2)

Enhancing Mitigation

4. The Task Force recommends that the NRC strengthen SBO mitigation capability at all operating and new reactors for design-basis and beyond-design-basis external events. (Section 4.2.1)

5. The Task Force recommends requiring reliable hardened vent designs in BWR facilities with Mark I and Mark II containments. (Section 4.2.2)

6. The Task Force recommends, as part of the longer term review, that the NRC identify insights about hydrogen control and mitigation inside containment or in other buildings as additional information is revealed through further study of the Fukushima Dai-ichi accident. (Section 4.2.3)

7. The Task Force recommends enhancing spent fuel pool makeup capability and instrumentation for the spent fuel pool. (Section 4.2.4)

8. The Task Force recommends strengthening and integrating onsite emergency response capabilities such as EOPs, SAMGs, and EDMGs. (Section 4.2.5)

Strengthening Emergency Preparedness

9. The Task Force recommends that the NRC require that facility emergency plans address prolonged SBO and multiunit events. (Section 4.3.1)

10. The Task Force recommends, as part of the longer term review, that the NRC pursue additional EP topics related to multiunit events and prolonged SBO. (Section 4.3.1)

11. The Task Force recommends, as part of the longer term review, that the NRC should pursue EP topics related to decisionmaking, radiation monitoring, and public education. (Section 4.3.2)

Improving the Efficiency of NRC Programs

12. The Task Force recommends that the NRC strengthen regulatory oversight of licensee safety performance (i.e., the ROP) by focusing more attention on defense-in-depth requirements consistent with the recommended defense-in-depth framework. (Section 5.1)

7. APPLICABILITY AND IMPLEMENTATION STRATEGY FOR NEW REACTORS

The Task Force has considered the applicability and implementation of its recommendations for new reactors, including certified designs, designs in the certification process, certified designs applying for renewal, early site permits, and applications for operating licenses and COLs. Since no new reactors are currently licensed, the recommendations involving orders are not applicable. However, the Task Force has assessed the rulemaking recommendations, and they would be equally applicable to new reactors.

Recommendation 8 for the integration of EOPs, SAMGs, and EDMGs and for controlling accident decisionmaking under technical specifications would be applicable to COLs. For near-term COLs (i.e., those expected to be licensed before the NRC completes the proposed rulemakings), the Task Force recommends that the agency impose those requirements through inspections, tests, analyses, and acceptance criteria (ITAAC). The Task Force recognizes that, with the sole exception of fire protection programs, the Commission has expressed (in the staff requirements memorandum dated September 11, 2002, for SECY-02-0067, "Inspections, Tests, Analyses, and Acceptance Criteria (ITAAC) for Operational Programs (Programmatic ITAAC)," dated April 5, 2002) a desire to resolve programmatic issues (e.g., training, quality assurance, fitness for duty) before COL issuance and handle the issues through existing oversight programs rather than resolve them through the ITAAC process. Specifically, the Commission stated (in the staff requirements memorandum for SECY-02-0067) that, "They [ITAAC] should encompass only those matters that, by their nature, cannot be resolved prior to construction." Further, the Commission stated, "...the Commission is not prepared to dismiss the possibility that programmatic ITAAC may be necessary in some very limited areas." The Task Force suggests that this would be one of those areas in which it is not practical to resolve the issue before COL issuance, in that the integration of EOPs, SAMGs, and EDMGs could require a few years of effort by licensees, the industry, and the NRC staff. However, this strategy would ensure implementation and NRC oversight before plant operation.

The Task Force concludes that all of the current early site permits already meet the requirements of detailed recommendation 2.1, relating to the design-basis seismic and flooding analysis, and all of the current COL and design certification applicants are addressing them adequately in the context of the updated state-of-the-art and regulatory guidance used by the staff in its reviews.

The Task Force concludes that Recommendation 4, with new requirements for prolonged SBO mitigation, and Recommendation 7, about spent fuel pool makeup capability and instrumentation, should apply to all design certifications or to COL applicants if the recommended requirements are not addressed in the referenced certified design. The Task Force recommends that design certifications and COLs under active staff review address this recommendation before licensing.

The Task Force notes that the two design certifications currently in the rulemaking process (i.e., the AP1000 and the economic simplified boiling-water reactor (ESBWR)) have passive safety systems. By nature of their passive designs and inherent 72-hour coping capability for core, containment, and spent fuel pool cooling with no operator action required, the ESBWR and AP1000 designs have many of the design features and attributes necessary to address the Task Force recommendations. The Task Force supports completing those design certification rulemaking activities without delay. However, COL applicants referencing these designs would have to address prestaging of any needed equipment for beyond 72 hours, and ITAAC should be established to confirm effective implementation of minimum and extended coping, as described in detailed recommendation 4.1.

Since the Task Force recommends the SBO additions on the basis of adequate protection, the NRC should impose them as new requirements in accordance with 10 CFR 52.59(b)(1) as part of the staff's review of the recently docketed Advanced Boiling-Water Reactor design certification renewal applications.

The recommendations related to expanding 10 CFR 50.54(hh) and the EP requirements to fully address multiunit accidents and SBO conditions should apply to COL applicants. Near-term COLs could implement these recommendations through ITAAC.

For the two plants with reactivated construction permits (Watts Bar Unit 2 and Bellefonte Unit 1), the Task Force recommends that those operating license reviews and the licensing itself include all of the near-term actions and any of the recommended rule changes that have been completed at the time of licensing. Any additional rule changes would be imposed on the plants in the same manner as for other operating reactors.

APPENDIX A

Summary of Detailed Recommendations by Implementation Strategy

In this appendix, the Task Force proposes an integrated strategy for implementing its recommendations in a coherent manner. The structure of this process includes (1) a policy statement, (2) rulemaking activities, (3) orders, (4) staff actions, and (5) actions for long-term evaluation.

As described in Section 3 of the report, in light of the low likelihood of an event beyond the design basis of a U.S. nuclear power plant and the current mitigation capabilities at those facilities, the Task Force concludes that continued operation of these plants and continued licensing activities do not pose an imminent risk to the public health and safety. Further, the Task Force concludes that the current regulatory approach and regulatory requirements continue to serve as a basis for the reasonable assurance of adequate protection of public health and safety until the actions set forth below have been implemented.

Policy Statement

Draft a Commission policy statement that articulates a risk-informed defense-in-depth framework that includes extended design-basis requirements in the NRC's regulations as essential elements for ensuring adequate protection. (Section 3—detailed recommendation 1.1)

Recommended Rulemaking Activities

The Task Force recommends that the Commission direct the staff to initiate these important rulemaking activities, including concurrent development of associated guidance, and complete them as a soon as possible.

- Initiate rulemaking to implement a risk-informed, defense-in-depth framework consistent with the above recommended Commission policy statement. (Section 3—detailed recommendation 1.2)

- Initiate rulemaking to require licensees to confirm seismic hazards and flooding hazards every 10 years and address any new and significant information. If necessary, update the design basis for SSCs important to safety to protect against the updated hazards. (Section 4.1.1—detailed recommendation 2.2)

- Initiate rulemaking to revise 10 CFR 50.63 to require each operating and new reactor licensee to (1) establish a minimum coping time of 8 hours for a loss of all ac power, (2) establish the equipment, procedures, and training necessary to implement an "extended loss of all ac" coping time of 72 hours for core and spent fuel pool cooling and for reactor coolant system and primary containment integrity as needed, and (3) preplan and prestage offsite resources to support uninterrupted core and spent fuel pool cooling, and reactor coolant system and containment integrity as needed, including the ability to deliver the equipment to the site in the time period allowed for extended coping, under conditions involving significant degradation of offsite transportation infrastructure associated with significant natural disasters. (Section 4.2.1—detailed recommendation 4.1)

- Initiate rulemaking or licensing activities or both to require the actions related to the spent fuel pool described in detailed recommendations 7.1–7.4. (Section 4.2.5—detailed recommendation 7.5)

- Initiate rulemaking to require more realistic, hands-on training and exercises on SAMGs and EDMGs for all staff expected to implement the strategies and those licensee staff expected to make decisions during emergencies, including emergency coordinators and emergency directors. (Section 4.2.6—detailed recommendation 8.4)

- Initiate rulemaking to require EP enhancements for multiunit events in the following areas: personnel and staffing, dose assessment capability, training and exercises, and equipment and facilities. (Section 4.3.1—detailed recommendation 9.1)

- Initiate rulemaking to require EP enhancements for prolonged SBO in the following areas: communications capability, ERDS capability, training and exercises, and equipment and facilities. (Section 4.3.1—detailed recommendation 9.2)

Recommended Orders

The Task Force recommends that the Commission use orders to ensure that licensees take the near-term actions described below. In some cases, these are interim actions to be taken until requirements associated with future rulemakings can be implemented.

- Order licensees to reevaluate the seismic and flooding hazards at their sites against current NRC requirements and guidance, and, if necessary, update the design basis and SSCs important to safety to protect against the updated hazards. (Section 4.1.1—detailed recommendation 2.1)

- Order licensees to perform seismic and flood protection walkdowns to identify and address plant-specific vulnerabilities and verify the adequacy of monitoring and maintenance for protection features such as watertight barriers and seals in the interim period until longer term actions are completed to update the design basis for external events. (Section 4.1.1—detailed recommendation 2.3)

- Order licensees to provide reasonable protection for equipment currently provided pursuant to 10 CFR 50.54(hh)(2) from the effects of design-basis external events and to add equipment as needed to address multiunit events while other requirements are being revised and implemented. (Section 4.2.1—detailed recommendation 4.2)

- Order licensees to include a reliable hardened vent in boiling-water reactor (BWR) Mark I and Mark II containments. (Section 4.2.3—detailed recommendation 5.1)

- Order licensees to provide sufficient safety-related instrumentation, able to withstand design-basis natural phenomena, to monitor key spent fuel pool parameters (i.e., water level, temperature, and area radiation levels) from the control room. (Section 4.2.5—detailed recommendation 7.1)

- Order licensees to provide safety-related ac electrical power for the spent fuel pool makeup system. (Section 4.2.5—detailed recommendation 7.2)

- Order licensees to revise their technical specifications to address requirements to have one train of onsite emergency electrical power operable for spent fuel pool makeup and spent fuel pool instrumentation when there is irradiated fuel in the spent fuel pool, regardless of the operational mode of the reactor. (Section 4.2.5—detailed recommendation 7.3)

- Order licensees to have an installed, seismically qualified means to spray water into the spent fuel pools, including an easily accessible connection to supply the water (e.g., using a portable pump or pumper truck) at grade outside the building. (Section 4.2.5—detailed recommendation 7.4)

- Order licensees to modify the EOP technical guidelines (required by Supplement 1, "Requirements for Emergency Response Capability," to NUREG-0737, issued January 1983 (GL 82-33), to (1) include EOPs, SAMGs, and EDMGs in an integrated manner, (2) specify clear command and control strategies for their implementation, and (3) stipulate appropriate qualification and training for those who make decisions during emergencies. (Section 4.2.6—detailed recommendation 8.1)

- Order licensees to modify each plant's technical specifications to conform with detailed recommendation 8.2. (Section 4.2.6—detailed recommendation 8.3)

- Order licensees to do the following until rulemaking is complete: determine and implement the required staff to fill all necessary positions for responding to a multiunit event, conduct periodic training and exercises for multiunit and prolonged SBO scenarios, ensure that EP equipment and facilities are sufficient for dealing with multiunit and prolonged SBO scenarios, provide a means to power communications equipment needed to communicate onsite and offsite during a prolonged SBO, and maintain ERDS capability throughout the accident. (Section 4.3.1—detailed recommendation 9.3)

- Order licensees to complete the ERDS modernization initiative by June 2012 to ensure multiunit site monitoring capability. (Section 4.3.1—detailed recommendation 9.4)

Recommended Staff Actions

The Task Force recommends that the staff begin the actions given below.

- Modify the Regulatory Analysis Guidelines to more effectively implement the defense-in-depth philosophy in balance with the current emphasis on risk-based guidelines. (Section 3—detailed recommendation 1.3)

- Evaluate the insights from the IPE and IPEEE efforts as summarized in NUREG1560, "Individual Plant Examination Program: Perspectives on Reactor Safety and Plant Performance," issued December 1997, and NUREG-1742, "Perspectives Gained from the Individual Plant Examination of External Events (IPEEE) Program," issued April 2002, to identify potential generic regulations or plant-specific regulatory requirements. (Section 3—detailed recommendation 1.4)

- Modify Section 5.0, "Administrative Controls," of the Standard Technical Specifications for each operating reactor design to reference the approved EOP technical guidelines for that plant design. (Section 4.2.6—detailed recommendation 8.2)

- Expand the scope of the annual ROP self assessment and biennial ROP realignment to more fully include defense-in-depth considerations. (Section 5.1—detailed recommendation 12.1)

- Enhance NRC staff training on severe accidents, including training resident inspectors on SAMGs. (Section 5.1—detailed recommendation 12.2)

Recommended Actions for Long-Term Evaluation

The Task Force recommends that the staff pursue the longer term review activities described below to further evaluate insights from the Fukushima event and to enhance the safety of U.S. plants.

- Evaluate potential enhancements to the capability to prevent or mitigate seismically induced fires and floods. (Section 4.1.2—detailed recommendation 3)

- Reevaluate the need for hardened vents for other containment designs, considering the insights from the Fukushima accident. Depending on the outcome of the reevaluation, appropriate regulatory action should be taken for any containment designs requiring hardened vents. (Section 4.1.3—detailed recommendation 5.2)

- Identify insights about hydrogen control and mitigation inside containment or in other buildings as additional information is revealed through further study of the Fukushima Dai-ichi event. (Section 4.1.4—detailed recommendation 6)

- Analyze current protective equipment requirements for emergency responders and guidance based upon insights from the accident at Fukushima. (Section 4.3.1—detailed recommendation 10.1)

- Evaluate the command and control structure and the qualifications of decisionmakers to ensure that the proper level of authority and oversight exists in the correct facility for a long-term SBO or multiunit accident or both. (Section 4.3.1—detailed recommendation 10.2)

- Evaluate ERDS to do the following: determine an alternate method (e.g., via satellite) to transmit ERDS data that does not rely on hardwired infrastructure that could be unavailable during a severe natural disaster, determine whether the data set currently being received from each site is sufficient for modern assessment needs, and determine whether ERDS should be required to transmit continuously so that no operator action is needed during an emergency. (Section 4.3.1—detailed recommendation 10.3)

- Study whether enhanced onsite emergency response resources are necessary to support the effective implementation of the licensees' emergency plans, including the ability to deliver the equipment to the site under conditions involving significant natural events where degradation of offsite infrastructure or competing priorities for response resources could delay or prevent the arrival of offsite aid. (Section 4.3.2—detailed recommendation 11.1)

- Work with FEMA, States, and other external stakeholders to evaluate insights from the implementation of EP at Fukushima to identify potential enhancements to the U.S. decisionmaking framework, including the concepts of recovery and reentry. (Section 4.3.2—detailed recommendation 11.2)

- Study the efficacy of real-time radiation monitoring onsite and within the EPZs (including consideration of ac independence and real-time availability on the Internet). (Section 4.3.2—detailed recommendation 11.3)

- Conduct training, in coordination with the appropriate Federal partners, on radiation, radiation safety, and the appropriate use of KI in the local community around each nuclear power plant. (Section 4.3.2—detailed recommendation 11.4)

APPENDIX B

Tasking Memorandum

March 23, 2011

MEMORANDUM TO: R. W. Borchardt
                    Executive Director for Operations

FROM: Chairman Jaczko   /RA/

SUBJECT: TASKING MEMORANDUM – COMGBJ-11-0002 – NRC
                    ACTIONS FOLLOWING THE EVENTS IN JAPAN

The staff should establish a senior level agency task force to conduct a methodical and systematic review of our processes and regulations to determine whether the agency should make additional improvements to our regulatory system and make recommendations to the Commission for its policy direction. The review should address the following near term and then longer term objectives.

Near Term Review

- This task force should evaluate currently available technical and operational information from the events that have occurred at the Fukushima Daiichi nuclear complex in Japan to identify potential or preliminary near term/immediate operational or regulatory issues affecting domestic operating reactors of all designs, including their spent fuel pools, in areas such as protection against earthquake, tsunami, flooding, hurricanes; station blackout and a degraded ability to restore power; severe accident mitigation; emergency preparedness; and combustible gas control.
- The task force should develop recommendations, as appropriate, for potential changes to inspection procedures and licensing review guidance, and recommend whether generic communications, orders, or other regulatory requirements are needed.
- The task force efforts should be informed by some stakeholder input but should be independent of industry efforts.
- The report would be released to the public per normal Commission processes (including its transmission to the Commission as a Notation Vote Paper).

To ensure the Commission is both kept informed of these efforts and called upon to resolve any policy recommendations that surface, the task force should, at a minimum, be prepared to brief the Commission on a 30 day quick look report; on the status of the ongoing near term review at approximately the 60 day point; and then on the 90 day culmination of the near term efforts. Additional specific subject matter briefings and additional voting items that request Commission policy direction may also be added during the Commission's agenda planning meetings.
    (EDO)   (SECY Suspense: 30, 60, & 90 days)

- The task force's longer term review should begin as soon as NRC has sufficient technical information from the events in Japan with the goal of no later than the completion of the 90 day near term report, and the task force should provide updates on the beginning of the longer term review at the 30 and 60 day status updates.
- This effort would include specific information on the sequence of events and the status of equipment during the duration of the event.
- The task force should evaluate all technical and policy issues related to the event to identify potential research, generic issues, changes to the reactor oversight process, rulemakings, and adjustments to the regulatory framework that should be conducted by NRC.
- The task force should evaluate potential interagency issues such as emergency preparedness.
- Applicability of the lessons learned to non-operating reactor and non-reactor facilities should also be explored.
- During the review, the task force should receive input from and interact with all key stakeholders.
- The task force should provide a report with recommendations, as appropriate, to the Commission within six months from the start of the evaluation for Commission policy direction.
- The report would be released to the public per normal Commission processes (including its transmission to the Commission as a Notation Vote Paper).
- Before beginning work on the longer term review, staff should provide the Commission with estimated resource impacts on other regulatory activities.
- The ACRS should review the report as issued in its final form and provide a letter report to the Commission.

(EDO)   (SECY Suspense: 9 months, if
                                                   needed)

cc: Commissioner Svinicki
        Commissioner Apostolakis
        Commissioner Magwood
        Commissioner Ostendorff
        OGC
        CFO
        OCA
        OPA
        Office Directors, Regions, ACRS, ASLBP (via E-Mail)
        PDR

APPENDIX C

Memo Transmitting Charter for the Nuclear Regulatory Commission Task Force to Conduct a Near-Term Evaluation of the Need for Agency Actions Following the Events Iin Japan

March 30, 2011

MEMORANDUM TO:	Martin J. Virgilio Deputy Executive Director for Reactor and Preparedness Programs Executive Director for Operations
	Charles L. Miller, Director Office of Federal and State Materials and Environmental Management Programs
FROM:	R. W. Borchardt */RA/* Executive Director for Operations
SUBJECT:	AGENCY TASK FORCE TO CONDUCT NEAR-TERM EVALUATION OF THE NEED FOR AGENCY ACTIONS FOLLOWING THE EVENTS IN JAPAN

On March 11th, 2011, Japan experienced a severe earthquake resulting in the shutdown of multiple reactors. It appears that the reactors' response to the earthquake went according to design. At the Fukushima Daiichi site, the earthquake caused the loss of normal AC power. In addition, it appears that the ensuing tsunami caused the loss of emergency AC power at the Fukushima Daiichi site. Subsequent events caused damage to fuel and radiological releases offsite.

The purpose of this memorandum is to task the Deputy Executive Director for Reactor and Preparedness Programs (DEDR) to convene an agency task force of U.S. Nuclear Regulatory (NRC) senior leaders and experts. The task force should conduct a methodical and systematic review of relevant NRC regulatory requirements, programs, and processes, and their implementation, to recommend whether the agency should make near-term improvements to our regulatory system. The task force should also identify a framework and topics for review and assessment for the longer-term effort.

Attached is a charter for the task force. The charter defines the objective, scope, coordination and communication, expected products, schedule, staffing, and Executive Director for Operations interface. The task force should update the Commission on the near-term review at approximately 30 and 60 days, and provide its observations, findings, and recommendations in the form of a written report and briefing at the completion of the near-term effort occurring at approximately 90 days.

The review should be conducted in accordance with Tasking Memorandum – COMGBJ-11-0002, "NRC Actions Following the Events in Japan."

Enclosure: As stated

CONTACT: Nathan T. Sanfilippo, OEDO
             301-415-3951

CHARTER FOR THE NUCLEAR REGULATORY COMMISSION TASK FORCE

TO CONDUCT A NEAR-TERM EVALUATION OF THE NEED FOR AGENCY ACTIONS

FOLLOWING THE EVENTS IN JAPAN

<u>Objective</u>

The objective of this task force is to conduct a methodical and systematic review of relevant NRC regulatory requirements, programs, and processes, and their implementation, to recommend whether the agency should make near-term improvements to our regulatory system. This task force will also identify a framework and topics for review and assessment for the longer-term effort.

<u>Scope</u>

The task force review will include the following:

a. A near-term review to:

- Evaluate currently available technical and operational information from the events that have occurred at the Fukushima Daiichi nuclear complex in Japan to identify potential or preliminary near-term/immediate operational or regulatory actions affecting domestic reactors of all designs, including their spent fuel pools. The task force will evaluate, at a minimum, the following technical issues and determine priority for further examination and potential agency action:
    - External event issues (e.g. seismic, flooding, fires, severe weather)
    - Station blackout
    - Severe accident measures (e.g., combustible gas control, emergency operating procedures, severe accident management guidelines)
    - 10 CFR 50.54 (hh)(2) which states, "Each licensee shall develop and implement guidance and strategies intended to maintain or restore core cooling, containment, and spent fuel pool cooling capabilities under the circumstances associated with loss of large areas of the plant due to explosions or fire, to include strategies in the following areas: (i) Fire fighting; (ii) Operations to mitigate fuel damage; and (iii) Actions to minimize radiological release." Also known as B.5.b.
    - Emergency preparedness (e.g., emergency communications, radiological protection, emergency planning zones, dose projections and modeling, protective actions)
- Develop recommendations, as appropriate, for potential changes to NRC's regulatory requirements, programs, and processes, and recommend whether generic communications, orders, or other regulatory actions are needed.

b. Recommendations for the content, structure, and estimated resource impact for the longer-term review.

<u>Coordination and Communications</u>

The near-term task force will:

- Solicit stakeholder input as appropriate, but remain independent of industry efforts.
- Coordinate and cooperate where applicable with other domestic and international efforts reviewing the events in Japan for additional insights.
- Provide recommendations to the Commission for any immediate policy issues identified prior to completion of the near-term review.
- Provide recommendations to program offices for any immediate actions not involving policy issues, prior to completion of the near-term review.
- Identify resource implications of near-term actions.
- Consider information gained from Temporary Instruction 2515/183, "Followup to the Fukushima Daiichi Nuclear Station Fuel Damage Events."
- Develop a communications plan.
- Update and brief internal stakeholders, as appropriate.

<u>Expected Product and Schedule</u>

The task force will provide its observations, conclusions, and recommendations in the form of a written report to the Deputy Executive Director for Reactor and Preparedness Programs at the completion of the 90-day near-term review.

During the development of its report, the task force will brief the Commission on the status of the review at approximately the 30- and 60-day points.

The report will be transmitted to the Commission via a SECY paper, and the task force will brief the Commission on the results of the near-term effort at approximately the 90-day point. The report will be released to the public via normal Commission processes.

The task force will recommend a framework for a longer-term review as a part of the near-term report. The longer-term review will begin as soon as the NRC has sufficient technical information from the events in Japan (with a goal of beginning by the end of the near-term review).

Staffing

The task force will consist of the following members:

Leader	Charles Miller	FSME
Senior Managers	Daniel Dorman	NMSS
	Jack Grobe	NRR
	Gary Holahan	NRO
Senior Staff	Amy Cubbage	NRO
	Nathan Sanfilippo	OEDO
Administrative Assistant	Cynthia Davidson	OGC

Additional task force members will be added as needed. For the near-term review, other staff members may be consulted on a part-time basis.

EDO Interface

The task force will keep agency leadership informed on the status of the effort and provide early identification of significant findings. The task force will report to Martin J. Virgilio, Deputy Executive Director for Reactor and Preparedness Programs.

This page intentionally left blank

www.ingramcontent.com/pod-product-compliance
Lightning Source LLC
Chambersburg PA
CBHW081830170526
45167CB00007B/2767